Performance Analysis of Mobile and Wireless Network Systems

Performance Analysis of Mobile and Wireless Network Systems

Edited by **Glynn Clermont**

WILLFORD PRESS
New York

Published by Willford Press,
118-35 Queens Blvd., Suite 400,
Forest Hills, NY 11375, USA
www.willfordpress.com

Performance Analysis of Mobile and Wireless Network Systems
Edited by Glynn Clermont

© 2016 Willford Press

International Standard Book Number: 978-1-68285-152-4 (Hardback)

Printed in the United States of America.

Contents

Preface

This book aims to highlight the current researches and provides a platform to further the scope of innovations in this area. This book is a product of the combined efforts of many researchers and scientists from different parts of the world. The objective of this book is to provide the readers with the latest information in the field.

Mobile and wireless network systems play a very crucial role in modern communication systems. This book is a compilation of various case-studies and researches that analyze the performance of different mobile and wireless networks. The topics covered in this book such as wireless multimedia systems, architectures and routing protocols for wireless and mobile networks, algorithms and communication channels, etc. provide a detailed explanation of the important concepts and applications of mobile and wireless network systems. This text will prove to be a very useful source of reference for students and professionals alike.

I would like to express my sincere thanks to the authors for their dedicated efforts in the completion of this book. I acknowledge the efforts of the publisher for providing constant support. Lastly, I would like to thank my family for their support in all academic endeavors.

Editor

Enhanced Gain and Bandwidth of Patch Antenna Using EBG Substrates

Mst. Nargis Aktar[1], Muhammad Shahin Uddin[2], Md. Ruhul Amin[3], and Md. Mortuza Ali[4]

[1]Department of Information and Communication Technology
Mawlana Bhashani Science and Technology University, Bangladesh
[2]Department of Electronics, Kookmin University, Seoul, South Korea
[3]Department of Electrical and Electronic Engineering
Islamic University of Technology, Dhaka, Bangladesh
[4]Department of Electrical and Electronic Engineering
Rajshahi University of Engineering and Technology, Bangladesh

E-mail: {nargismbstu@gmail.com, shahin.mbstu@gmail.com,
aminr_bd@yahoo.com, and mmali.ruet@gmail.com}

ABSTRACT

Microstrip patch antenna becomes very popular day by day because of its ease of analysis and fabrication, low cost, light weight, easy to feed and their attractive radiation characteristics. Although patch antenna has numerous advantages, it has also some drawbacks such as restricted bandwidth, low gain and a potential decrease in radiation pattern. In recent years, attention to use Electromagnetic Band Gap (EBG) substrates to overcome the limitations of patch antenna. In this paper, we propose a rectangular microstrip patch antenna with EBG substrates and compare the performance of the proposed antenna with a conventional patch antenna in the same physical dimension. Due to the presence of the EBG structure in the dielectric substrates, the electromagnetic band gap is created that reduces the surface waves considerably. As a result, the performance of the proposed antenna is better comparing the conventional existing microstrip patch antenna.

KEYWORDS

Microstrip patch antenna, Electromagnetic band gap (EBG) substrates, Gain and Bandwidth.

1. INTRODUCTION

With the drastic demand of wireless communication system and their miniaturization, antenna design becomes more challenging. Recently microstrip patch antennas have been widely used in satellite communications, aerospace, radars, biomedical applications and reflector feeds because of its inherent characteristics such as light weight, low profile, low cost, mechanically robust, compatibility with integrated circuits and very versatile in terms of resonant frequency, polarization, pattern and impedance . In spite of its several advantages, they suffer from drawbacks such as narrow bandwidth, low gain and excitation of surface waves, etc [1-3]. These drawbacks limit their applications in other fields. In order to overcome the limitations of microstrip patch antennas such as narrow bandwidth and low gain, numerous techniques are proposed i.e. for probe fed stacked antenna, microstrip patch antennas on electrically thick substrate, slotted patch antenna and stacked shorted patches have been proposed and investigated [4-5]. These methods have eliminated the bandwidth problem for most applications. But limitations of gain and surface wave excitation still remain. That is why, in recent years there has been considerable effort in the EBG structure for antenna application to suppress the surface wave and overcome the limitations of the antenna. Many works have been done to improve the performance of the microstrip antennas [6-11]. The EBG structure utilizes the inherent properties of dielectric materials to enhance the microstrip antenna performance. EBG materials are periodic dielectrics that produce pass band and stop band characteristics.

In this paper, we propose a rectangular patch antenna with EBG substrates. The characteristics of EBG depend on the shape, size, symmetry and the material used in their construction. Surface waves are reduced by using EBG substrate which leads to increase the directivity, bandwidth and radiation efficiency [15]. EBG were realized to reduce and eliminate surface waves, which leads to an increase in directivity, bandwidth and radiation efficiency. It is also useful to reduce the side lobes of the radiation pattern and hence radiation pattern front-to-back ratio and overall antenna efficiency are improved. Our proposed antenna gives better performance compare to the conventional rectangular microstrip patch antenna. A substantial gain and bandwidth enhancement has been obtained. The design and simulation have been done by using High Frequency Structure Simulator (HFSS). The remainder of the paper is organized as follows: In section II, a brief description of EBG structure. In section III present the conventional and proposed antenna design and configuration. In section IV present the simulation results and discussion. The conclusion of this paper is provided in section V.

2.   ELECTROMAGNETIC BAND GAP SUBSTRATES

The birth of the electromagnetic band gap structure has triggered many novel antenna applications. Electromagnetic band gap structures can be defined as artificial periodic (or sometimes non-periodic) objects that prevent or assist the propagation of electromagnetic waves in a specified band of frequency for all incident angle and polarization state. Two commonly employed features are suppressing unwanted substrate modes and acting as an artificial magnetic ground plane. The main advantage of EBG structure is their ability to suppress the surface wave current. The generation of surface waves degrades the antenna efficiency and radiation pattern. Furthermore, it increases the mutual coupling of the antenna array which causes the blind angle of a scanning array [12-14].

EBG structures are usually realized by periodic arrangement of dielectric materials and metallic conductors. In general, they can be categorized into three groups according to their geometric configuration; (i) three-dimensional volumetric structures, (ii) two-dimensional planar surfaces, and (iii) one-dimensional transmission lines [13]. Two-dimensional planar EBG surfaces again classified into two categories, first one is mushroom like EBG surfaces and another one is uniplanar EBG surfaces.

For the mushroom like EBG surfaces, a band gap is observed between the frequency 7GHz and 11GHz. On the other hand, for the uniplanar EBG surfaces a band gap is observed the frequency from 13GHz to 14.6GHz. In this paper, mushroom like EBG surface is used in order to design patch antenna on EBG substrates because the mushroom like EBG surface has a lower frequency band gap and a wider bandwidth than the uniplanar EBG surface.

A two dimensional mushroom like EBG structure is shown in Figure 1. Design of patch antenna mushroom like EBG structures are preferable because light weight, low fabrication cost. There are four main parameters affecting the performance of mushroom like EBG structures. The parameters are like this: rectangle width w, gap width g, substrates thickness h and substrates permittivity ε_r. Also, the vertical vias radius r has a trival effect because it is very thin compared to the operating wavelength. The parameters that are affecting the performance of EBG structures are directly dependent on the operating wavelength of the patch antenna [8]. The parameters are varying with operating wavelength as like this that the rectangle width, w varies from $0.04\lambda_{12\ GHz}$ to $0.20\lambda_{12\ GHz}$, gap width varies from $0.01\lambda_{12\ GHz}$ to $0.12\lambda_{12\ GHz}$ and the substrate thickness, h varies from $0.01\lambda_{12\ GHz}$ to $0.09\lambda_{12\ GHz}$. Here, λ_{12} means the wavelength between medium 1 and 2 i.e. the free space and the guiding device and GHz means the wavelength respect to the GHz range frequency.

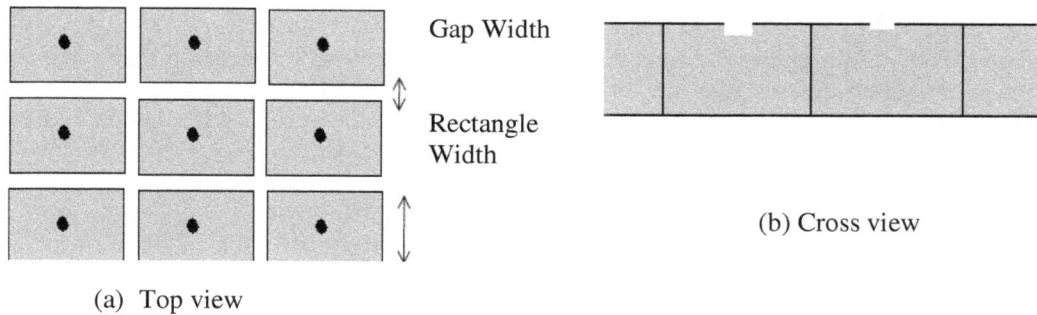

Figure 1 Two dimensional mushrooms like EBG surfaces: (a) Top view (b) Cross view

3.   ANTENNA DESIGN AND CONFIGURATION

In order to identify and verify the improvement of the performance of microstrip antenna on EBG substrates, designed a conventional antenna and the proposed antenna. The width of the rectangular patch antenna is usually chosen to be larger than the length of the patch, L to get higher bandwidth. The antenna is designed to operate at frequency 10GHz. In this paper, we use neltec dielectric material as patch substrates whose dielectric constant is 2.45. The antenna is excited by a microstrip transmission line feed. The point of excitation is adjustable to control the impedance match between feed and antenna, polarization, mode of operation and excitation frequency. To design patch antenna lower dielectric constant is used because in case of lower dielectric constant of the substrates, surface wave losses are more severe and dielectric and conductor losses are less severe. By using EBG structures, surface wave loss can be reduced easily. Table1 shows the important parameters for the geometrical configuration of the patch antenna.

Table1 Geometrical configuration of the patch antenna

Antenna Part	Parameter	Value
Patch	Length	8.8mm
	Width	11.4mm
Patch Substrates (NeltecNx9245) (IM)(tm)	Dielectric constant	2.45
	Height	0.787mm
	Dielectric loss tangent	0.01
EBG Substrates	Rectangle Width	$0.10\lambda_{12\,GHz}$
	Gap Width	$0.02\lambda_{12\,GHz}$
	Substrates thickness	$0.04\lambda_{12\,GHz}$
Operating Frequency		10GHz

4. SIMULATION RESULTS AND DISCUSSIONS

Now a days, it is a common practice to evaluate the system performances through computer simulation before the real time implementation. A simulator "Ansoft HFSS" based on finite-element method (FEM) has been used to calculate return loss, impedance bandwidth, radiation pattern and gains. This simulator also helps to reduce the fabrication cost because only the antenna with the best performance would be fabricated. Figure 2 shows the simulated results of the return loss of the conventional antenna and the proposed antenna.

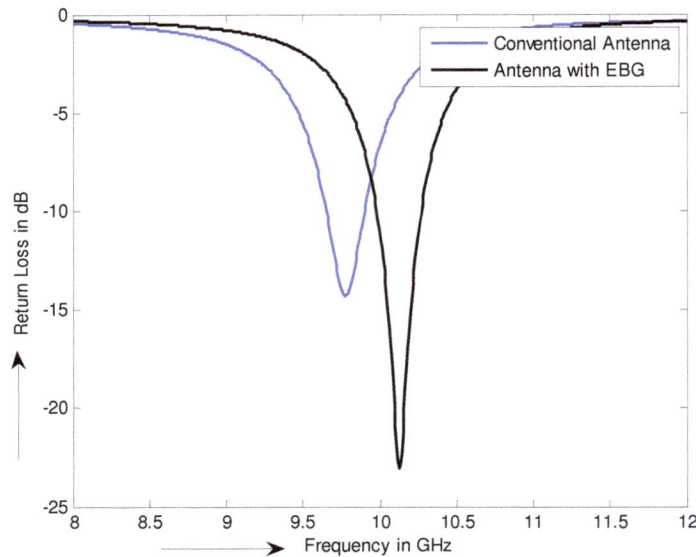

Figure 2 Return losses of the conventional patch antenna and antenna with EBG

It is seen from the Figure 2, the return loss for the conventional patch antenna is − 14.5dB at 9.79GHz and for the proposed patch antenna is -23dB at 10.12GHz. A negative value for return loss shows that this antenna had not many losses while transmitting the signals. According to theoretical design, the minimum loss has been observed at 10 GHz. But from simulation results we have observed that the minimum loss get at 9.79 GHz for conventional antenna and 10.12GHz for the proposed antenna. Thus the return loss of the proposed microstrip patch antenna is 58.6% less compared to the conventional microstrip patch antenna. From the same Figure 2, the antenna bandwidth can be calculated. At the point of return loss -10dB, the bandwidth (BW) and the relative bandwidth (RBW) are 240MHz and 5.43% for conventional patch antenna but at same point of return loss the bandwidth (BW) and the relative bandwidth (RBW) of the proposed antenna are 330MHz and 7.33%. Therefore, the bandwidth of the proposed antenna is 37.5% more than the conventional antenna.

The simulated results for gain that are obtained from conventional antenna and the proposed antenna on EBG substrates are shown in Figure 3 and Figure 4.

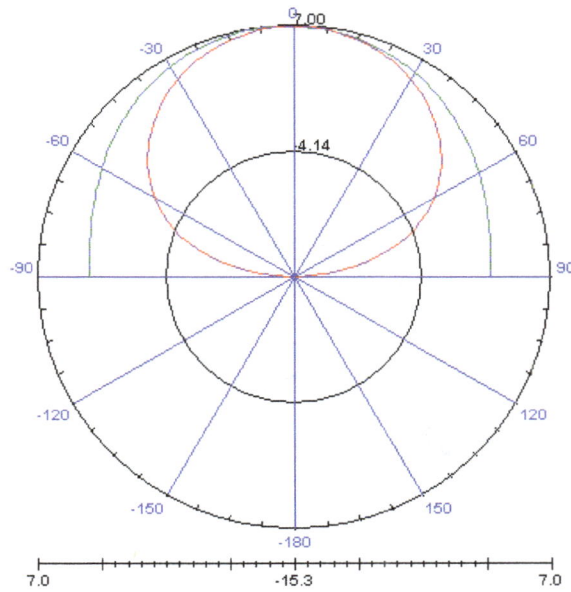

Figure 3 Gain of the conventional rectangular patch antenna.

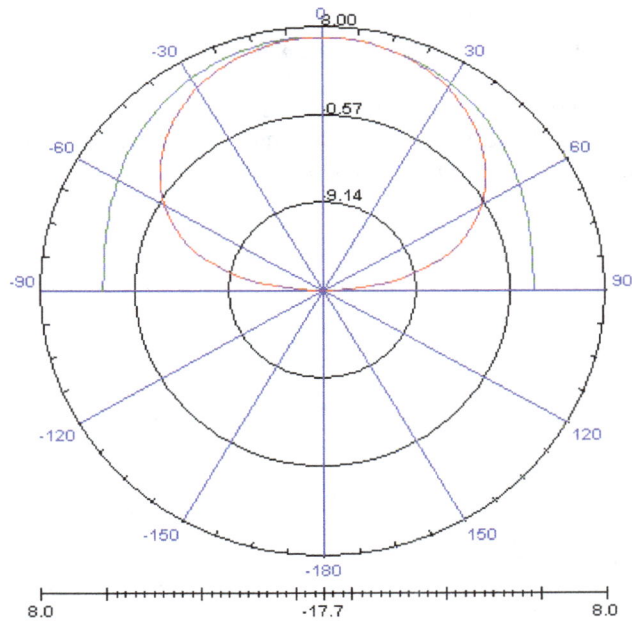

Figure 4 Gain of the rectangular patch antenna with EBG

From the simulated results, it is shown that the gain of the conventional antenna and the proposed antenna is 22.3dB and 25.7dB. So, the gain of the proposed patch antenna on EBG substrates is 15.2% more than the conventional patch antenna.

The Figure 5 and Figure 6 shows the simulated directivity of the conventional antenna and the proposed antenna.

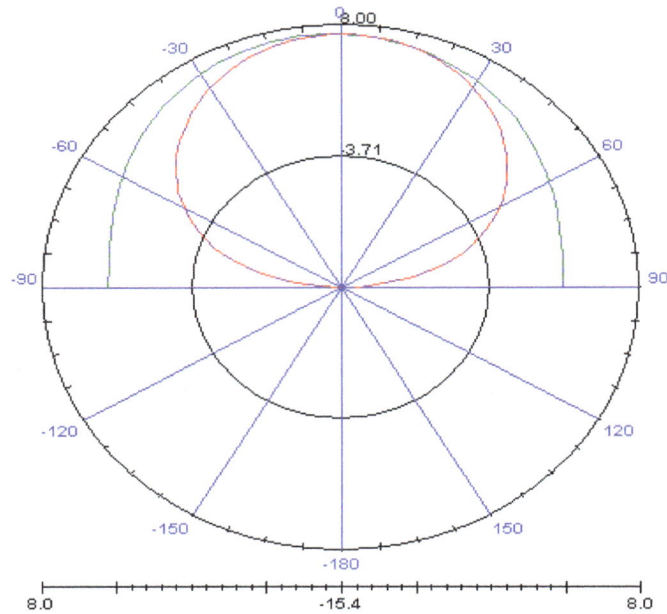

Figure 5 Directivity of the conventional rectangular patch antenna.

From the Figure 5 and Figure 6, the directivity for the conventional patch antenna and the proposed patch antenna are 23.4dBi and 25.5dBi. Thus, the directivity of the proposed antenna is also enhanced of 8.97% than the conventional antenna.

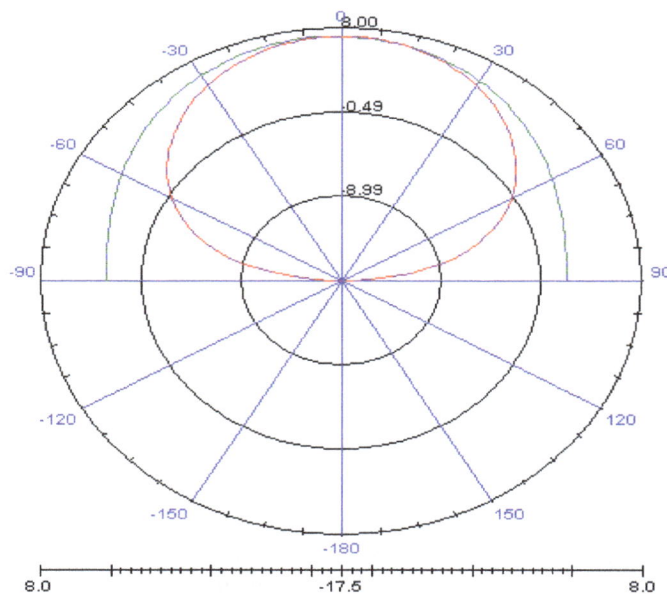

Figure 6 Directivity of the rectangular patch antenna with EBG

5. CONCLUSION

The patch antenna mostly used in modern mobile communication. The goals of this paper are to design conventional patch antenna and the patch antenna on EBG substrates with same physical dimensions that can operate at 10GHz and study the performance of misrostrip antenna when EBG structure added on it. From the simulated results, it is seen that the performance is better of a patch antenna that is designed on EBG substrates than the conventional patch antenna. In future, our targets are to real time implementation of the proposed antenna and also design another microstrip patch antenna with EBG substrates that can operate at higher frequency.

References

[1]   Jing Liang, and Hung-Yu David Yang, "Radiation Characteristics of a Microstrip Patch over an Electromagnetic Bandgap Surface," IEEE Transactions on Antennas and Propagation, Vol. 55, June 2007, pp1691-1697.

[2]   Mohammad Tariqul Islam, Mohammed Nazmus Shakib, Norbahiah Misran, and Baharudin Yatim, "Analysis of Broadband Slotted Microstrip Patch Antenna," Proceedings of the International Conference on Computer and Information Technology, December 2008, pp. 758-761.

[3]   K.L. Wong, Compact and Broadband Microstrip Antennas. New York: Wiley, 2002.

[4]   D. N. Elsheakh, IEEE, H. A. Elsadek, E. A. Abdallah, H. Elhenawy, and M. F. Iskander,"Enhancement of Microstrip Monopole Antenna Bandwidth by Using EBG Structures," IEEE Antennas and Wireless Propagation Letters, vol. 8, 2009, pp 959-962.

[5]   S. Pioch and J.M. Laheurte, "Low Profile Dual-Band Antenna Based on a Stacked Configuration of EBG and Plain Patches," Microwave. Opt. Tech. Letter, vol. 44, February 2005, pp. 207–209.

[6]   D. Qu, L. Shafai and A. Foroozesh, "Improving Microstrip Patch Antenna Performance Using EBG Substrates," IEE Proc. Micro. Antennas Propagation, Vol.153, December 2006, pp.558-563.

[7]   G. S. Kliros, K. S. Liantzas, A. A. Konstantinidis, "Modeling of Microstrip Patch Antennas with Electromagnetic Band Gap Superstrates," 19th International Conference on Applied Electromagnetics and Communications, 2007.

[8]   Ram´on Gonzalo, Peter de Maagt, and Mario Sorolla, "Enhanced Patch-Antenna Performance by Suppressing Surface Waves Using Photonic-Bandgap Substrates," IEEE Transactions on Microwave Theory and Techniques, Vol. 47, November 1999, pp. 2131-2138.

[9]   Atsuya Ando, Kenichi Kagoshima, Akira Kondo, and Shuji Kubota, "Novel Microstrip Antenna With Rotatable Patch Fed by Coaxial Line for Personal Handy-Phone System Units," IEEE Transactions on Antennas and Propagation, Vol. 56, August 2008, pp.2747-2751.

[10]   Mr. Pramod Kumar.M , Sravan kumar, Rajeev Jyoti , VSK Reddy, PNS Rao,"Novel Structural Design for Compact and Broadband Patch Antenna," IEEE 2010.

[11]   R. Chantalat, C. Menudier, M. Thevenot, T. Monediere, E. Arnaud, and P. Dumon,"Enhanced EBG Resonator Antenna as Feed of a Reflector Antenna in the Ka Band," IEEE Antennas and Wireless Propagation Letters, Vol. 7, 2008 pp.349-353

[12]   Nasimuddin, Zhi Ning Chen, Terence S. P. See, and Xianming Qing, "Multi-dielectric Layer Multi-Patches Microstrip Antenna for UWB Applications," Proceedings of the 37th European Microwave Conference, Munich Germany, October 2007, pp 1019-1021

[13]   Fan Yang, Yahya Rahmat Sami, "Electromagnetic band Gap Structures in Antenna Engineering," Cambridge University Press 2009.

[14]   Constantine A. Balanis, " Antenna Theory Analysis and Design," Third Edition, John Wiley & Sons, 2005

[15]   D. M. Pozar, "Microwave Engineering," Third edition. New York, Wiley, 2005.

2

Peak to Average Power Ratio Reduction of OFDM Signal by combining Clipping with Walsh Hadamard Transform

Navneet Kaur and Lavish Kansal

Department of Electronics and Communication Engineering
Lovely Professional University, Phagwara
Punjab, India
nv_neet@yahoo.com
Department of Electronics and Communication Engineering
Lovely Professional University, Phagwara
Punjab, India
lavish.15911@lpu.co.in

ABSTRACT

Wireless communications have been developed widely and rapidly in the modern world especially during the last decade. Orthogonal Frequency Division Multiplexing (OFDM) has grown to a popular communication technique for high speed communication. Besides of the advantages, one of main disadvantage of OFDM is high peak to average power ratio (PAPR). In this paper, a PAPR reduction method is proposed that is based on combining clipping with Walsh Hadamard Transform (WHT). WHT is a precoding technique which is having less complexity compared to the other existing power reduction techniques and also it can reduce PAPR considerably and results in no distortion. The performance of the proposed scheme is examined through computer simulations and it is found that power reductions are obtained.

KEYWORDS

OFDM, PAPR, WHT, MC, DMT, FFT, PSK, QAM SLM, PTS, CCDF.

1. INTRODUCTION

Recent advances in wireless communication systems have increased the throughput over wireless channels. The reliability of wireless communication has also been increased. But still the bandwidth and spectral availability demands are endless. The need to achieve reliable wireless systems with high spectral efficiency, low complexity and good error performance results in continued research in this field. To provide such a high spectral efficiency, an efficient modulation scheme is to be employed [1, 2]. A promising modulation technique that is increasingly being considered is Orthogonal Frequency Division Multiplexing (OFDM) due to its advantages in dealing with the multipath propagation problem, high data rate and bandwidth efficiency [3].Being an important member of the multicarrier modulation (MC) techniques, OFDM is also called Discrete Multitone Modulation (DMT). It is based upon the principle of frequency division multiplexing (FDM) where each frequency channel is modulated with simpler modulation scheme. It splits a high rate data stream into a number of lower rate streams that are transmitted simultaneously over a number of orthogonal subcarriers [4]. Orthogonality is achieved by ensuring that the carriers are placed exactly at the nulls in the modulation spectra of each other. The increase of symbol duration for the lower rate parallel subcarriers reduces the relative amount of dispersion in time caused by multipath delay spread.

However, OFDM also has its shortcoming. The major drawback of OFDM signal is its large peak-to-average power ratio (PAPR), which causes poor power efficiency or serious performance degradation to transmit power amplifier. To reduce the PAPR, many techniques have been proposed. Such as clipping, partial transmit sequence (PTS) [5], selected mapping (SLM) [6], interleaving, nonlinear companding transforms, hadamard transforms [7] and other techniques etc. [8, 9]. These schemes can mainly be categorized into signal scrambling techniques, such as PTS, and signal distortion techniques such as clipping. Among those PAPR reduction methods, the simplest scheme is to use the clipping process.

The organization of this paper is as follow. Section 2 presents OFDM signal model and formulates the problem of PAPR. Section 3, Walsh Hadamard Transform precoding is discussed. Existing clipping technique is described in section 4.Proposed scheme based on combined WHT transform and clipping technique is explained in section 5. Section 6 presents simulation results. Conclusions are drawn in section 7.

2. OFDM SIGNAL MODEL AND PAPR PROBLEM

2.1 OFDM System

The simulation model OFDM system is presented in figure 1. This model consists of a transmitter, a channel and a receiver.

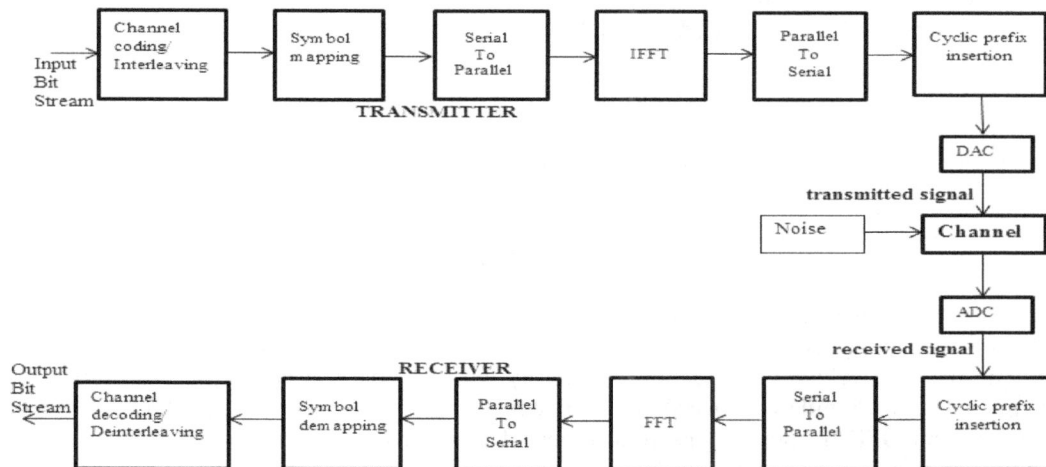

Figure 1. Block Diagram of OFDM system

In an OFDM scheme, a large number of orthogonal, overlapping, narrow band sub-carriers are transmitted in parallel each sub-carrier is then modulated with a conventional modulation scheme such as Quadrature Amplitude Modulation (QAM) or Phase-Shift Keying (PSK) at a low symbol rate than that required for the whole data stream, but still maintaining total data rates similar to conventional single-carrier modulation schemes in the same bandwidth. A high-rate data stream is split into a number of lower rate streams to be transmitted simultaneously over a number of sub-carriers. Since the symbol duration increases for lower rate parallel sub-carriers, the amount of dispersion in time caused due to multipath delay is reduced. These carriers divide the available transmission bandwidth. The separation of the sub-carriers is such that there is a very compact spectral utilization and each being modulated at a low bit rate. In a conventional frequency division multiplex the carriers are individually filtered to ensure there is no spectral overlap.

2.2 PAPR in OFDM System

OFDM signal is expressed as a sum of independent subcarriers in the time and frequency directions and multiplied by the data symbols. An OFDM signal in baseband is defined as:

$$s(t) = \sum_{n=0}^{N-1} \left(b_n e^{j2\pi f_n t} w(t) \right); \ 0 \leq t \leq T \tag{1}$$

where, b_n denotes the complex symbol modulating the *n-th* carrier, *w(t)* is the time window function defined in the interval [0,T], N is the number of subcarriers, and T is the duration of an OFDM symbol. Subcarriers are spaced $\Delta f = 1/T$ apart. Each subcarrier is located at:

$$f = \frac{n}{T}; 0 < n < N\text{-}1 \tag{2}$$

In order to maintain the orthogonality between the OFDM symbols, the symbol duration and sub channel space must meet the condition $T\Delta f = 1$. Presence of large number of independently modulated subcarriers in an OFDM system the peak value of the system can be very high as compared to the average of the whole system. This ratio of the peak to average power value is termed as Peak-to-Average Power Ratio. Coherent addition of N signals of same phase produces a peak which is N times the average signal. High PAPR increases the complexity in the analog to digital and digital to analog converter and reduces the efficiency of RF amplifiers. In OFDM, a block of N symbols $\{b_n, n = 0,1, \dots, N - 1\}$ is formed with each symbol modulating one of a set of subcarriers; $\{f, n = 0,1, \dots, N - 1\}^n$. The resulting signal is given as:

$$s\,(t) = \frac{1}{\sqrt{N}} \sum_{n=0}^{N-1} b_n e^{j2\pi\Delta f t}; \ 0 \leq t \leq NT \tag{3}$$

The PAPR of the transmitted signal is defined as:

$$PAPR = \frac{max|s(t)|^2}{\frac{1}{NT} \int_0^{NT} |s(t)|^2 dt} \tag{4}$$

Reducing the *max|s (t)|* is the principle goal of PARP reduction techniques. The PAPR reduction capability is measured by the empirical complementary cumulative distributive function (CCDF), which indicates the probability that the PAPR is above a certain threshold. Complementary Cumulative Distribution Function (CCDF) curves present vital information regarding the OFDM signal to be transmitted. The main use of power CCDF curves is to identify the power characteristic of the signals which are amplified, mixed and decoded [10] i.e. to find the probability of the PAPR. The ratio between power level and the average power is expressed in db. For a transmitted *s (t)* OFDM signal CCDF of $\{PAPR\{s(t) > \gamma\}\}$ is given as:

$$Prob(PAPR\{s(t)\} > \gamma) = 1 - (1 - e^{-\gamma})^N \tag{5}$$

where N is the number of subcarriers.

3. CLIPPING TECHNIQUE

The clipping [8] technique employs clipping around the peaks to reduce the PAPR. It is simple to implement. It reduces the PAPR by simply limiting the maximum amplitude of the OFDM signal, such that all signal values are limited to the threshold. Clipping the OFDM signal before amplification is a simple method to limit PAPR. The clipping operation is carried out at the transmitter. The clipping operation on the real band pass signal is given by:

$$k(t) = \begin{cases} -A ; & if \ s(t) \leq -A \\ s(t); & if -A \leq s(t) \leq A \\ A; & if \ s(t) \geq A \end{cases} \qquad (6)$$

where $k(t)$ is the clipped signal and A is the clipping level. After which the exceeded signal was clipped.

Figure 2(a) shows the plot of an OFDM signal having maximum amplitude of 0.8.The peak of the OFDM signal exceeds the value from 0.6.This exceeded peak of the signal causes PAPR.

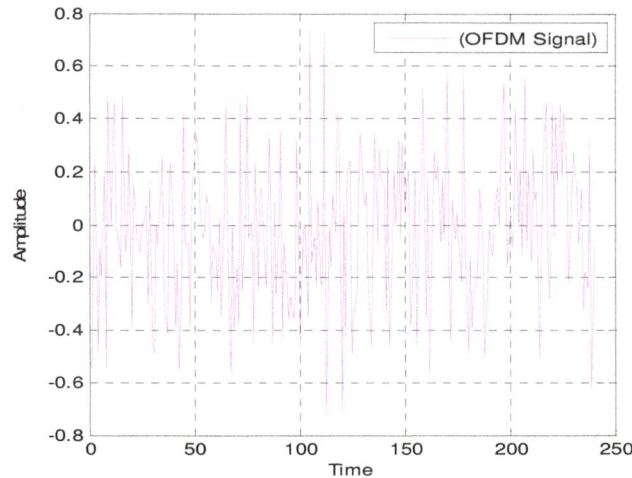

Figure 2(a). OFDM Signal with high peak value

To reduce the peak a threshold value was assumed in Figure 2(b) the value of the signal exceeding the value of 0.5 and -0.5 was clipped off. This reduction in the peak value limits the PAPR in the system.

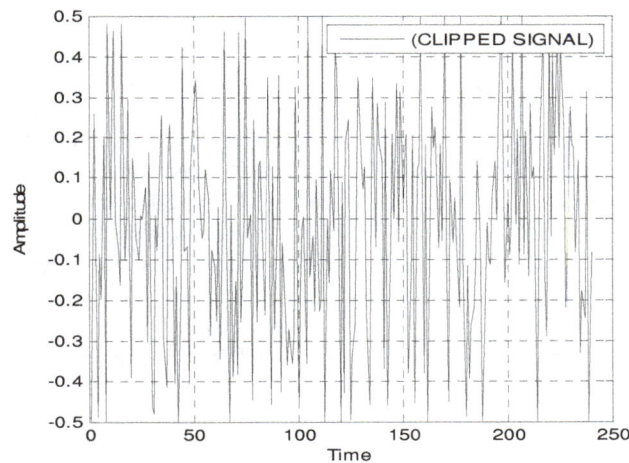

Figure 2(b).OFDM Signal with clipped peak value

4. WALSH HADAMARD TRANSFORM TECHNIQUE

The goal of precoding techniques is to obtain a signal with lower PAPR than in the case of OFDM without precoding techniques and to reduce the interference produced by multiple users. The PAPR reduction must compensate the non linearities of the HPA having as effect the reduction of the bit error rate (BER). The main characteristics of precoding based techniques are: no bandwidth expansion, no power increase, and no data rate loss, no BER degradation and distortion less. WHT precoding technique is presented in the following.

The technique of Hadamard Transform [7] is based upon the relationship between correlation property of OFDM input sequence and PAPR probability. Theaverage power of the input sequence represents the peak value of the autocorrelation. Hence the peak value of autocorrelation depends on the input sequence provided thatnumber of sub carriers remains unchanged.

The Walsh Hadamard Transform (WHT) is a non-sinusoidal, orthogonal linear transform and can be implemented by a butterfly structure as in FFT. This means that applying WHT does not require the extensive increase of system complexity. WHT decomposes a signal into set of basic functions. These functions are Walsh functions, which are square waves with values of +1 or -1 [11]. The proposed hadamard transform scheme may reduce the occurrence of the high peaks comparing the original OFDM system. The idea to use the WHT is to reduce the autocorrelation of the input sequence to reduce the peak to average power problem and it requires no side information to be transmitted to the receiver. The kernel of WHT can be written as follows:

$$H_1 = [1]$$

$$H_2 = \frac{1}{\sqrt{2}} \begin{bmatrix} 1 & 1 \\ 1 & -1 \end{bmatrix}$$

$$H_{2N} = \frac{1}{\sqrt{2N}} \begin{bmatrix} H_N & H_N \\ H_N & H_N^{-1} \end{bmatrix} \tag{7}$$

where H_N^{-1} denotes the binary complement of H_N.between each.

5. PROPOSED SCHEME

In the proposed scheme two appropriate methods are combined to reduce PAPR. One is the existing technique called clipping and the other is the precoding technique called walsh hadamard transform. The proposed combined peak clipping is different than normal peak clipping method because in normal peak clipping only one peak of OFDM symbol is clipped, whereas in proposed combined method PAPR reduces to much extent than by simple clipping The pre-coding has been considered as a best among all these techniques because it improves PAPR without increasing much complexity and without destroying the orthogonality between subcarriers. The combined techniques based OFDM system was shown in fig. 3. In proposed scheme conventional clipping was processed after WHT technique. At the transmitter end, the data stream is firstly transformed by WHT, and then the transformed data is processed by the clipping unit. If data block passed by WHT, before IFFT, the PAPR is reduced, then the PAPR of OFDM signal could be further reduced by clipping.

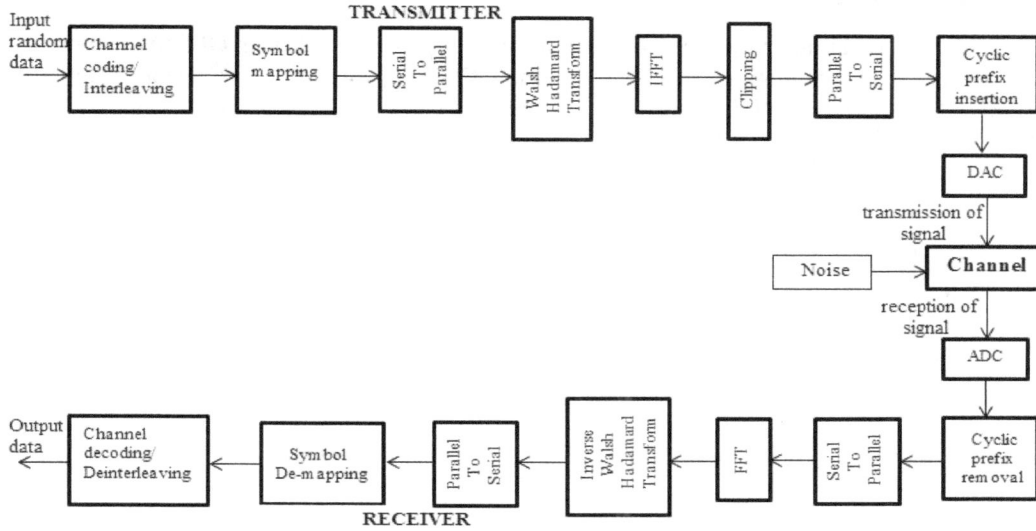

Figure 3. Block scheme of WHT Precoding technique with clipping in OFDM system

WHT precoding based OFDM system was shown in fig. 3. In these system, the kernel of the WHT acts as a precoding matrix K of dimension $P=L \times L$ and it is applied to constellations symbols before the IFFT to reduce the correlation among the input sequence. In the precoding based systems baseband modulated data is passed through S/P converter which generates a complex vector of size L that can be written as $D=[D_0, D_1, ..., D_{L-1}]^T$. Then precoding is applied to this complex vector which transforms this complex vector into new vector of length L that can be written as $b_n = KD = [b_0, b_1, ..., b_{L-1}]^T$ where P is a precoder matrix of size $N=L \times L$ and b_n can be written as follows:

$$b_n = \sum_{l=0}^{L-1} K_{m,l}.D_l \quad ; m=0,\ 1,...,\ L\text{-}1 \tag{8}$$

$K_{m,l}$ means m^{th} row and l^{th} column of precoder matrix. The complex baseband OFDM signal with N subcarriers can be written as:

$$s(t) = \frac{1}{\sqrt{N}} \sum_{m=0}^{N-1} b_n.e^{j2\pi m \frac{n}{N}};\ n=0,\ 1,\ 2,...,\ N\text{-}1 \tag{9}$$

Clipping is performed on the I and Q outputs of the IFFT after WHT precoding was applied on OFDM signal. As the word length at the IFFT output is decreased, the power consumption and complexity of the DAC/ADC decreases. Also clipping at the IFFT output increases the resolution giving a better average signal.

6. SIMULATION RESULTS

In this section, the PAPR of OFDM with WHT precoding technique has been evaluated by simulation. To show PAPR analysis of the proposed system, the data is generated randomly then the signal is modulated by M-PSK and M-QAM respectively. The block implementation is shown in Fig. 3, where the precoding matrix transform represents proposed walsh hadamard transform precoding technique used in our simulations. We can evaluate the performance of the PAPR reduction scheme using the complementary cumulative distribution ($CCDF= Prob(PAPR\{s(t)\} > \gamma$) of the PAPR of the OFDM signal. The CCDF of the PAPR for WHT precoded OFDM signal is used to express the probability of exceeding a given threshold. We compared the simulation results of proposed system with WHT precoded OFDM systems and conventional OFDM systems.

6.1 M-PSK Modulation

In this section WHT technique combined with clipping is applied over OFDM system with phase shift keying modulation (PSK).We assume CCDF clip rate of $10^{-0.7}$ using M-ary PSK technique for subcarriers N=2400 and for different values of M results were evaluated where M=16, 32, 64, 128, 256, 512 and 1024.

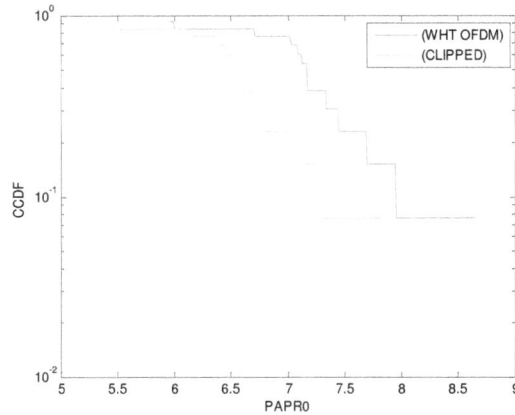

Figure 4. CCDF of clipping with proposed WHT technique for 16 PSK

For M=16 for PSK modulation, the CCDF performance of the proposed clipped scheme compared with that of the WHT precoded technique for OFDM signal was shown in **Figure 4**.At CCDF clip rate of $10^{-0.7}$ the PAPR value of clipped precoded signal reduces by 0.75 dB over WHT precoded system.

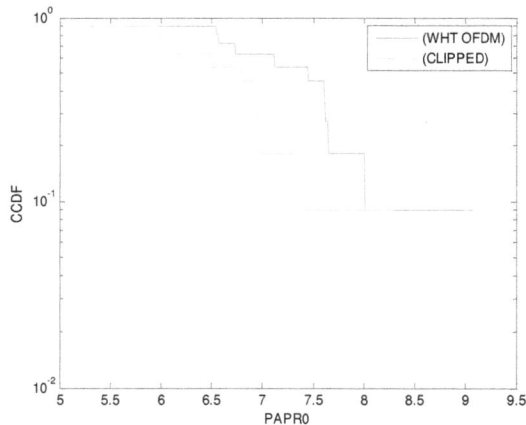

Figure 5. CCDF of clipping with proposed WHT technique for 32 PSK

For M=32 for PSK modulation, the CCDF performance of the proposed clipped scheme compared with that of the WHT precoded technique for OFDM signal was shown in **Figure 5**. At CCDF clip rate of $10^{-0.7}$ the PAPR value of clipped precoded signal reduces by 0.72 dB over WHT precoded system.

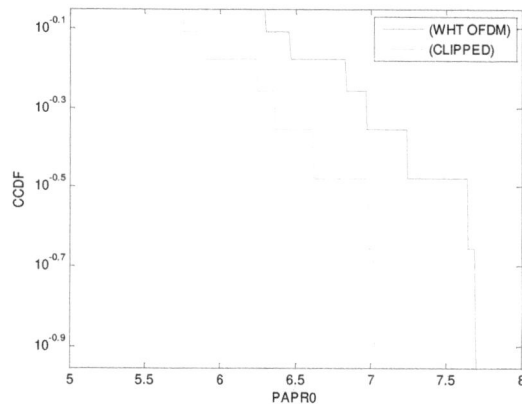

Figure 6. CCDF of clipping with proposed WHT technique for 64 PSK

For M=64 for PSK modulation, the CCDF performance of the proposed clipped scheme compared with that of the WHT precoded technique for OFDM signal was shown in **Figure 6**. At CCDF clip rate of $10^{-0.7}$ the PAPR value of clipped precoded signal reduces by 0.69 dB over WHT precoded system.

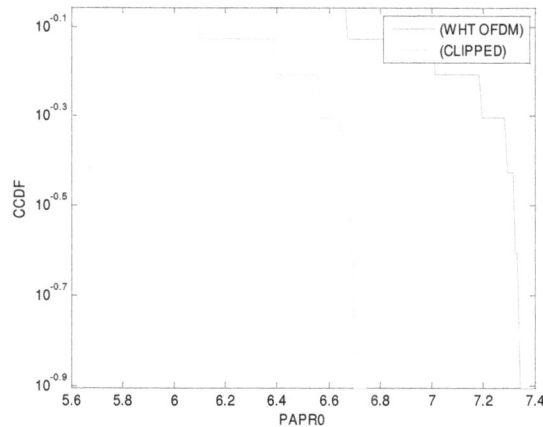

Figure 7. CCDF of clipping with proposed WHT technique for 128 PSK

For M=128 for PSK modulation, the CCDF performance of the proposed clipped scheme compared with that of the WHT precoded technique for OFDM signal was shown in **Figure 7**. At CCDF clip rate of $10^{-0.7}$ the PAPR value of clipped precoded signal reduces by 0.67 dB over WHT precoded system.

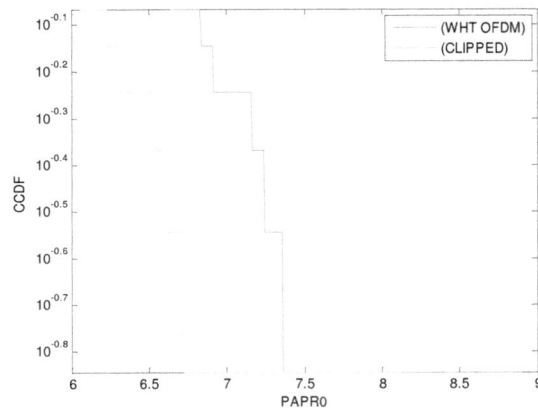

Figure 8. CCDF of clipping with proposed WHT technique for 256 PSK

For M=256 for PSK modulation, the CCDF performance of the proposed clipped scheme compared with that of the WHT precoded technique for OFDM signal was shown in **Figure 8**. At CCDF clip rate of $10^{-0.7}$ the PAPR value of clipped precoded signal reduces by 0.64 dB over WHT precoded system.

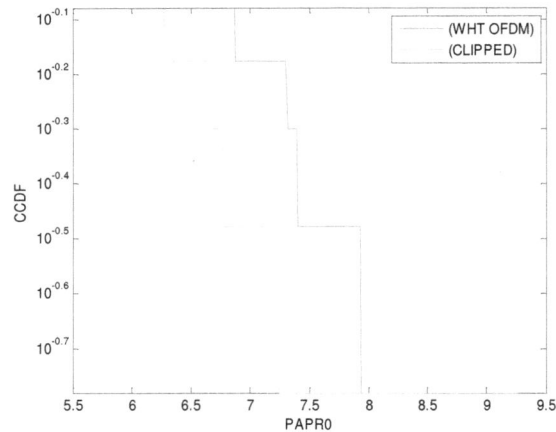

Figure 9. CCDF of clipping with proposed WHT technique for 512 PSK

For M=512 for PSK modulation, the CCDF performance of the proposed clipped scheme compared with that of the WHT precoded technique for OFDM signal was shown in **Figure 9**. At CCDF clip rate of $10^{-0.7}$ the PAPR value of clipped precoded signal reduces by 0.62 dB over WHT precoded system.

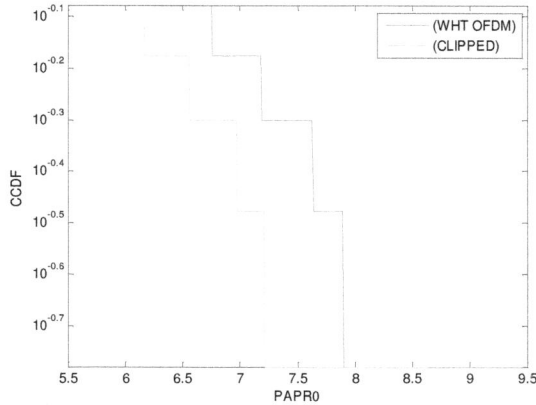

Figure 10. CCDF of clipping with proposed WHT technique for 1024 PSK

For M=1024 for PSK modulation, the CCDF performance of the proposed clipped scheme compared with that of the WHT precoded technique for OFDM signal was shown in **Figure 10**. At CCDF clip rate of $10^{-0.7}$ the PAPR value of clipped precoded signal reduces by 0.61 dB over WHT precoded system.

5.2 M-QAM Modulation

In this section using M-ary quadrature amplitude modulation (QAM) technique is applied for N=2400 and where M=16, 64,256 and 1024.

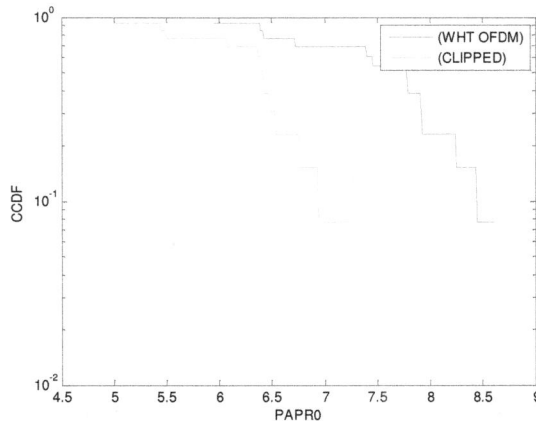

Figure 11. CCDF of clipping with proposed WHT technique for 16 QAM

For M=16 for QAM modulation, the CCDF performance of the proposed clipped scheme compared with that of the WHT precoded technique for OFDM signal was shown in **Figure 11**. At CCDF clip rate of $10^{-0.7}$ the PAPR value of clipped precoded signal reduces by 1.6 dB over WHT precoded system.

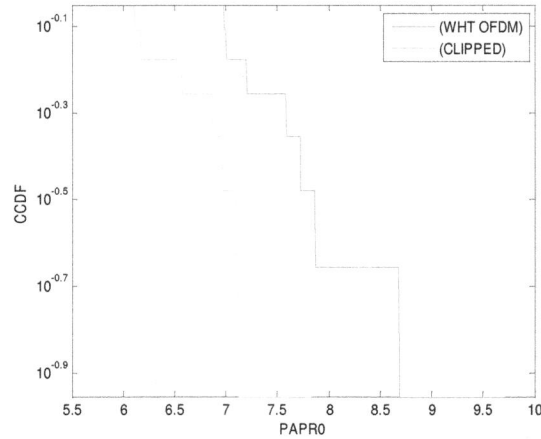

Fig. 12: CCDF of clipping with proposed WHT technique for 64 QAM

For M=64 for QAM modulation, the CCDF performance of the proposed clipped scheme compared with that of the WHT precoded technique for OFDM signal was shown in **Figure 12**. At CCDF clip rate of $10^{-0.7}$ the PAPR value of clipped precoded signal reduces by 1.52 dB over WHT precoded system.

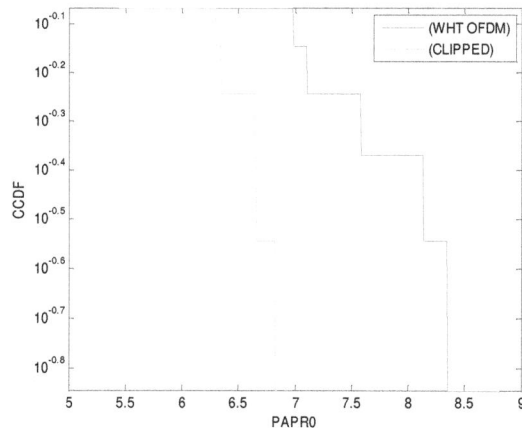

Fig. 13: CCDF of clipping with proposed WHT technique for 256 QAM

For M=256 for QAM modulation, the CCDF performance of the proposed clipped scheme compared with that of the WHT precoded technique for OFDM signal was shown in **Figure 13**. At CCDF clip rate of $10^{-0.7}$ the PAPR value of clipped precoded signal reduces by 1.5 dB over WHT precoded system.

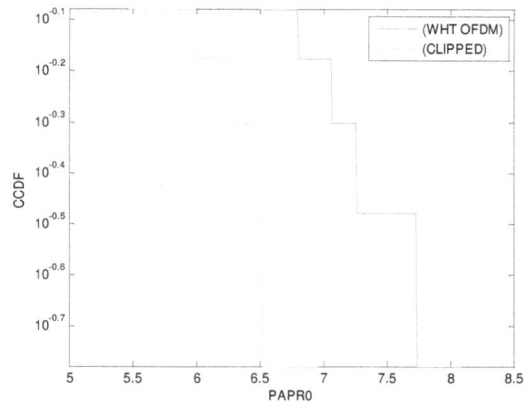

Fig. 14: CCDF of clipping with proposed WHT technique for 1024 QAM

For M=1024 for QAM modulation, the CCDF performance of the proposed clipped scheme compared with that of the WHT precoded technique for OFDM signal was shown in **Figure 14**. At CCDF clip rate of $10^{-0.7}$ the PAPR value of clipped precoded signal reduces by 1.3 dB over WHT precoded system.

7. CONCLUSIONS

In this paper, to reduce PAPR in OFDM system, we evaluated performance of clipping with WHT precoding technique. The proposed combined technique is simple to implement and has no limitations on the system parameters such as number of subcarriers modulation order, and constellation type. This system produce the lowest PAPR and is efficient, signal independent, distortion less and do not require any complex optimizations representing better PAPR reduction methods than others existing techniques because it does not require any power increment, complex optimization and side information to be sent to the receiver.

From simulation results, it can be observed that the proposed combined clipped method displays a better PAPR reduction performance than the WHT precoded OFDM signal. Thus, it is concluded that the proposed combined clipped scheme is more favourable than the precoded WHT transform.

ACKNOWLEDGEMENTS

Foremost, I would like to express my sincere gratitude to **Mr Lavish Kansal** who gave his full support in the compilation of this report with his stimulating suggestions and encouragement to go ahead in all the time of the thesis. At last but not the least my gratitude towards my parents, I would also like to thank God for the strength that keep me standing and for the hope that keep me believing that this report would be possible.

REFERENCES

[1] R.W Chang, "Synthesis of Band-Limited Orthogonal Signals for Multi-channel Data Transmission," Bell System Technology, Vol.45, pp.1775-1797, 1966.

[2] R.W Chang, "Orthogonal Frequency Division Multiplexing," U.S Patent 3388455,1966.

[3] S .Weinstein &P. Ebert, "Data Transmission by Frequency Division Multiplexing Using the Discrete Fourier Transform," IEEE Transaction on Communication, Vol.19, Issue: 5, pp. 628-634, 1971.

[4]   B.R. Satzberg, "Performance of an Efficient Parallel Data Transmission System,"IEEE Transaction on Communication, Vol. 15, Issue: 6, pp. 805-811, 1967.

[5]   R. W. Bauml, R. F. Fischer & J. B. Huber, "Reducing the Peak-to-Average Power Ration of Multicarrier Modulation by Selected Mapping," Electronics Letters, Vol. 32,Issue: 22, pp. 2050-2057, 1996.

[6]   S. H. Muller & J. B. Huber, "OFDM with Reduced Peak to Average Power Ratio by Optimum Combination of Partial Transmit Sequences," IEEE Electronics Letters, Vol. 33, Issue5, pp. 368-369, 1997.

[7]   M. Park, J. Heeyong, J. Cho, N. Cho, D. Hong & C. Kang, "PAPR Reduction in OFDM Transmission Using Hadamard Transform," IEEE International Conference on Communications, Vol. 1, pp. 430-433, 2000.

[8]   T. Jiang & Y. Wu, "An Overview: Peak to Average Power Ratio Reduction Techniques for OFDM Signals," IEEE Transactions on Broadcasting, Vol. 54, Issue: 2, pp. 257-268, 2008.

[9]   S. H. Han &J. H. Lee, "An Overview of Peak-to-Average Power Ratio Reduction Techniques for Multicarrier Transmission," IEEE Transactions on Wireless Communications, Vol. 12,Issue: 2, April 2005, pp. 56-65, 2005.

[10] Agilent Technologies Application Note, "Characterizing Digitally Modulated Signal with CCDF Curves", 2000.

[11] H. Rohling, Broadband OFDM Radio Transmission for Multimedia Applications, In IEEE Proceeding on Vehicular Conference, Vol. 87, Issue: 10,   pp.1778-1788, 1999.

[12] L. Kansal, A. Kansal & K. Singh, "Analysis of Different High Level Modulation Techniques for OFDM System," International Journal of VLSI and Signal Processing Applications, Vol. 1, Issue: 2, 2011.

[13] L. Kansal, A. Kansal & K. Singh, "Performance Analysis of MIMO-OFDM System Using QOSTBC Code Structure for M-PSK," Signal Processing: An International Journal, Vol. 5, Issue: 2, 2011.

[14] N. Kaur& L. Kansal, "Reducing the Peak to Average Power Ratio of OFDM Signals through Walsh Hadamard Transform," Global Journal of Researches in Engineering, Vol. 13, Issue: 1, 2013.

3

Performance Comparison of MIMO Systems over AWGN and Rician Channels with Zero Forcing Receivers

Navjot Kaur and Lavish Kansal

Lovely Professional University, Phagwara,
E-mails: er.navjot21@gmail.com, lavish.15911@lpu.co.in

Abstract

Multiple-Input Multiple-Output (MIMO) systems have been emerged as a technical breakthrough for high-data-rate wireless transmission. The performance of MIMO system can be improved by using different antenna selection so as to provide spatial diversity. In this paper, the performance of MIMO system over AWGN (Additive White Gaussian Noise) and Rician fading channels with ZF receiver is analyzed using different antenna configurations. The bit error rate performance characteristics of Zero-Forcing (ZF) receiver is studied for M-PSK (M-ary Phase Shift Keying) modulation technique using AWGN and Rician channels for the analysis purpose and their effect on BER (Bit Error Rate) have been presented.

Keywords – MIMO, spatial diversity, AWGN, Rician, fading, ZF, antenna, BER, M-PSK.

I. INTRODUCTION

MIMO systems make use of multiple antennas at the transmitter and receiver so as to increase the data rates by means of spatial diversity. So MIMO systems are well-known in wireless communications for high data rates. [1] The capacity of wireless systems can be increased by varying the number of antennas.

The two primary reasons for using wireless communication over wired communication:
- First is multi-path fading i.e. the variation of the signal strengths due to the various obstacles like buildings, path loss due to attenuation and shadowing [2].
- Second, for the wireless users, the transmission media is air as compared to the wired communication where each transmitter–receiver pair is considered as an isolated point-to-point link.

MIMO system utilizes the feature of spatial diversity by using spatial antennas in a dense multipath fading environment which are separated by some distance [3]. MIMO systems are implemented to obtain diversity gain or capacity gain to avoid signal fading. The idea to improve the link quality (BER) or data rate (bps) is the basic consideration behind the development of MIMO systems by using multiple TX/RX antennas [4]. The core scheme of MIMO is space-time coding (STC). The two main functions of STC: diversity & multiplexing. The maximum performance needs tradeoffs between diversity and multiplexing.

MIMO system employs various coding techniques for multiple antenna transmissions have become one of the desirable means in order to obtain high data rates over wireless channels [5]. However, of considerable concern is the increased complexity incurred in the implementation of such systems. MIMO antenna systems are used in recent wireless communications like WiMAX, IEEE 802.11n and 3GPP LTE etc.

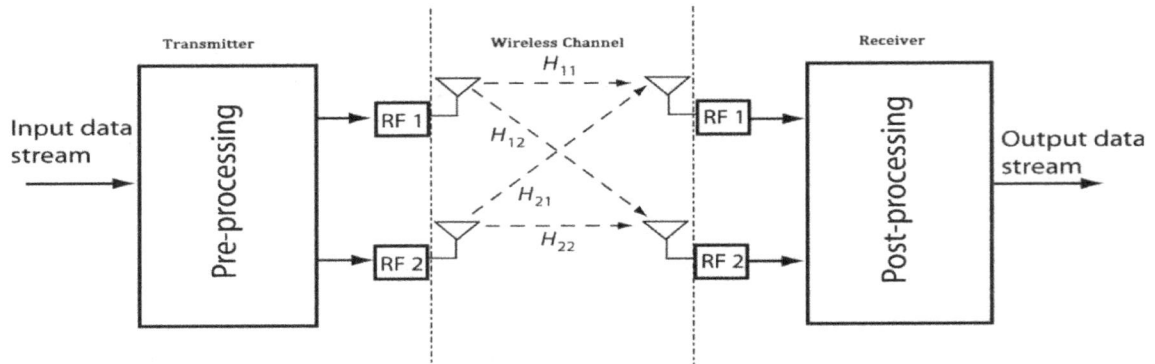

Fig. 1.1: MIMO System (2X2 MIMO Channel)

A. I. Sulyman [6] describes the performance of MIMO systems over nonlinear fading channels. The effects of antenna selection on its performance are also considered. The author has derived expressions for the PWEP performance of space-time trellis coding nonlinear Rayleigh fading channel. With the variation in the antenna selection at the receiver side, the performance degradation due to nonlinear fading channel reduces.

The comparison of MIMO with conventional Single-Input Single-Output (SISO) technology was discussed by S. G. Kim et. al [7]. The authors discussed that the MIMO system enhances the link throughput and also improves the spectral efficiency. The authors analyzed the BER performance of MIMO systems for M-PSK using ZF receiver over various fading channels in the presence of practical channel estimation errors.

C. Wang [8] explains the approach to increase the capacity of MIMO systems by employing spatial multiplexing. Maximum likelihood (ML) receiver achieves optimal performance whereas the linear receivers like Zero-Forcing (ZF) receiver provide sub-optimal performance. But Zero- Forcing receiver also offers significant reduction in computational complexity with performance degradation in tolerable limits.

A simple transmit diversity scheme comprises of two transmit antennas and one receive antenna was presented by X. Zhang et. al [9]. It provides the same spatial diversity order as that can be achieved by maximal-ratio receiver combining (MRRC) which makes use of one transmit antenna and two receive antennas.

A. Lozano et. al [10] compared the transmit diversity vs. spatial multiplexing in modern MIMO systems. Antenna diversity is a preferred weapon used by mobile wireless systems against the effect of fading. The prevalence of MIMO has opened the door for a much more effective use of antennas: spatial multiplexing.

The rich-scattering wireless channel is capable of enormous theoretical capacities if the multipath is properly exploited as per the researches done in the field of Information theory. P.

W. Wolniansky et. al [11], described an architecture of wireless communication known as V-BLAST (Vertical Bell Laboratories Layered Space-Time) that has been implemented in real-time environment.

An efficient implementation of space-time coding for the broadband wireless communications is presented by R. S. Blum et. al [12]. The authors presented the improved performance of MIMO-OFDM systems and diversity gains of a space time (ST) coding system through the type of trellis codes used in non-linear fading channel environment. The developed simulator for predicting the performance of a space time (ST) coded MIMO-OFDM system under different trellis coding and channel conditions is demonstrated.

The performance analysis of the low-cost effective MIMO system that employs the spatial multiplexing at the transmitter and zero-forcing processing at the receiver in multiuser scheduling systems was discussed by C. Chen [13]. By incorporating the mathematical tool of order statistics, the author derived the PDFs of effective sub channel output SNRs for a variety of scheduling algorithms. These expressions are used to derive the closed-form formulas. The closed-form expressions allow efficient numerical evaluations to characterize the capacity gain of this suboptimal transmission strategy under a number of practical scheduling policies requiring scalar or vector feedback. The results validate the elegant marriage of the zero-forcing receiver and scheduling technique as an economical approach to achieve higher data rates for next-generation wireless communications.

N. S. Kumar et. al [14], investigated about the three types of equalizer for MIMO wireless receivers. The authors discussed about a fixed antenna MIMO antenna configuration and compare the performance with all the three types of equalizer based receiver namely ZF, ML, and MMSE. BER performance of ML Equalizer is superior to zero forcing Equalizer and Minimum Mean Square Equalizers. It is inferred that the ML equalizer is the best of the three equalizers based on the mathematical modeling and the simulation results.

In this paper, the performance analysis of MIMO systems over AWGN and Rician channels using ZF receivers are presented. AWGN channel is a channel which has flat frequency response. It is known as universal channel model used for analyzing modulation schemes. In this, channel adds a white Gaussian noise to the signal passing through it. When there is line of sight, direct path is normally the strongest component goes into deeper fade compared to the multipath components. This kind of signal is approximated by Rician distribution.

II. BENEFITS OF MIMO SYSTEMS

Spatial multiplexing

Spatial multiplexing which comprises of number of transmit-receive antenna pairs tend to increase the transmission rate (or capacity) for the same bandwidth without any additional power expenditure. The increase in the transmission rate is proportional to the number of transmit-receive antenna pairs.

Interference reduction and avoidance

Multiple users which shares time and frequency resources result in interference in wireless networks. Interference may be mitigated in MIMO systems by exploiting the spatial dimension to increase the separation between users. To improve the coverage and range of a wireless network, there is need of interference reduction and avoidance.

Array gain

The coherent combining effect of multiple transmitting and receiving antennas tends to achieve good array gain at the receiver. This average increase in the SNR at the receiver requires perfect channel knowledge either at the transmitter or receiver or both.

Diversity gain

Multipath fading is the most significant problem in wireless communications due to various obstacles like building, scattering, reflection etc. In a fading channel, signal experiences fade (i.e the fluctuation in the signal strength). The channel is in deep fade when there is a significant drop in the signal power that gives rise to high BER. The diversity is used to so as to combat fading as much as it can.

Table 1.1: Benefits of MIMO system

MIMO TECHNIQUE	BENEFITS	BEST CONDITIONS
SPATIAL MULTIPLEXING	Increases the throughput of the system	Best performance is achieved at low velocity near to the base station (strong signal)
TRANSMIT DIVERSITY	Increases the range by countering fading (less possibility of errors) usually at base station	Good when beam forming is not appropriate
RECEIVE DIVERSITY	Increases the range by countering fading (less possibility of errors) usually at mobile station	Advantage over single antenna under all conditions
BEAMFORMING	Increases the range at base station	Works best at relatively low velocity when distance is extremely large (cell edge).

III. MODULATION TECHNIQUE

Modulation is the process of superimposing a low frequency information signal over a high frequency carrier signal so that its transmission is possible over a long distance. Modulation can be analog and digital type. Digital modulation maps the digital information over analog carrier so as to transmit it over the channel. Every digital communication system has a modulator in the transmitter side and a demodulator in the receiver side. Every transmitter has a modulator that performs the task of modulation. Every r

eceiver has a demodulator to perform the inverse process of modulation, called demodulation, so as to recover the transmitted digital information.

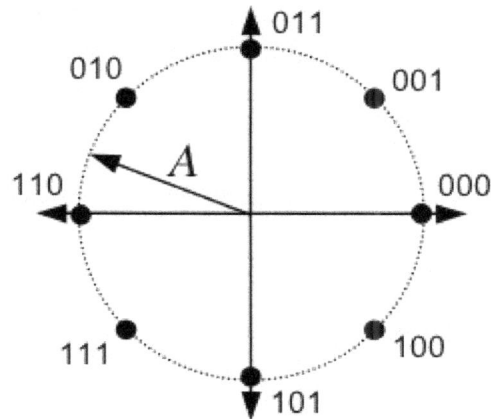

Fig. 1.2: Signal Space Diagram for 8-PSK

The M-ary PSK modulation yields circular constellation as the amplitude of the transmitted signals remains constant as shown in Fig. 1.2.

The signal set for M-ary Phase-shift keying (M-PSK) can be represented as:

$$X_i(t) = \frac{\sqrt{2E_s}}{T_s} \cos\left(2\pi * f_{c\tau} + \frac{2(i-1)}{M}\right) \quad i = 1,2,\dots\dots M \ \& \ 0 < \ < T_s \qquad (1.1)$$

where E_s represents the signal energy per symbol, T_s represents the symbol duration and $f_{c\tau}$ represents the carrier frequency.
This phase of the carrier changes for different possible values of M as follows:

$$\theta = 2(i-1)^{\pi/M} \quad i = 1,2,\dots\dots M \qquad (1.2)$$

IV. CHANNELS USED

Communication channels can be classified as fast and slow fading channels. In a fast channel, the impulse response changes approximately at the symbol rate of the communication system, whereas in a slow fading channel, it does not changes so frequently. Rather it stays unchanged for several symbols. In this paper, the performance analysis of MIMO system is discussed over the AWGN channel and Rician channel.

- AWGN channel: It is a channel used for analyzing modulation schemes by adding a white Gaussian noise to the signal passing through it. This channel's amplitude frequency response is flat and phase frequency response is linear for all frequencies. The modulated signals pass through it without any amplitude loss and phase distortion. So in such a case, fading does not exist but the only distortion that exists is introduced by the AWGN. The received signal is simplified to

$$r(t) = x(t) + n(t) \tag{1.3}$$

where n(t) represents the noise.

• Rician channel: When there is line of sight, direct path is normally the strongest component goes into deeper fade compared to the multipath components. This kind of signal is approximated by Rician distribution. As the dominating component run into more fade the signal characteristic goes from Rician to Rayleigh distribution. The signal characteristic goes from Rician to Rayleigh distribution as the dominating component run into more fade in multi-path fading.

$$p(r) = \frac{r}{\sigma^2} e^{-\frac{(r^2 + A^2)}{2\sigma^2}} I\left(\frac{Ar}{\sigma^2}\right) \quad \text{for } (A \geq 0, r \geq 0) \tag{1.4}$$

Where A denotes the peak amplitude (value) of the dominant signal and $I_0[.]$ is the modified Bessel function of zero-order.

V. MIMO SYSTEM MODEL

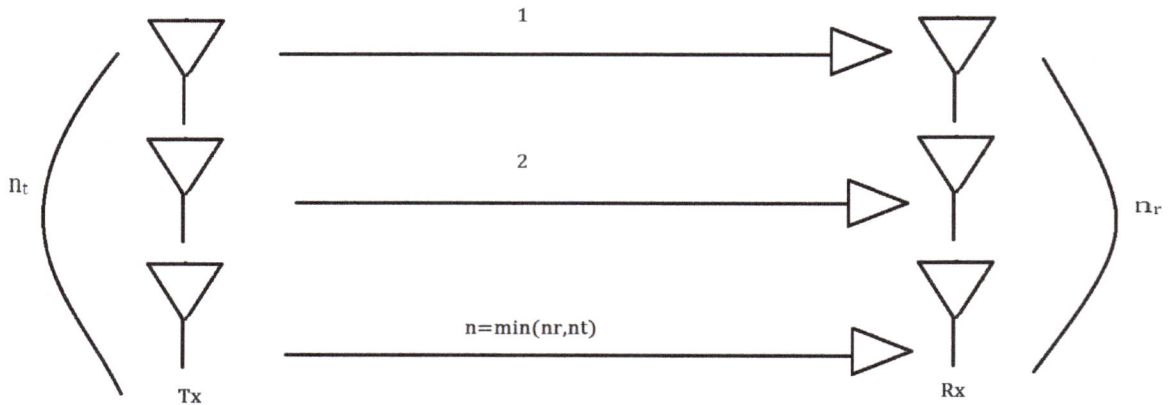

Fig. 1.3: MIMO channel as n SISO sub-channels

The MIMO channel is represented in Fig. 1.3 with an antenna array with n_t elements at the transmitter and an antenna array with n_r elements at the receiver is considered. The impulse response of the channel is $h_{ij}(\tau,t)$ between the jth transmitter element and the ith receiver element. The MIMO channel can then be described by the n_r X n_t $H(\tau,t)$ matrix:

$$H(\tau, t) = \begin{bmatrix} h_{1,1}(\tau, t) & h_{1,2}(\tau, t) & \cdots & h_{1,n_t}(\tau, t) \\ h_{2,1}(\tau, t) & h_{2,2}(\tau, t) & \cdots & h_{2,n_t}(\tau, t) \\ \vdots & \vdots & & \vdots \\ h_{n_r,n_1}(\tau, t) & h_{n_r,n_2}(\tau, t) & h_{n_r,n_t}(\tau, t) \end{bmatrix} \tag{1.5}$$

The matrix elements are complex numbers. These elements have dependency on the attenuation and phase shift that the wireless channel introduces delay τ to the received signal reaching at the receiver.

The input-output relation of the MIMO system can be expressed as follows:

$$y(t) = H(\tau, t) \otimes s(t) + u(t) \tag{1.6}$$

where $\otimes$ denotes convolution, $s(t)$ is a $n_t \times 1$ vector corresponding to the n_t transmitted signals, $y(t)$ is a $n_r \times 1$ vector corresponding to the n_r and $u(t)$ is the additive white noise.

VI. ZERO FORCING EQUALIZER

Zero Forcing Equalizer was first proposed by Robert Lucky, is a linear receiver used in communication systems. This equalizer inverts the frequency response of the channel to the received signal so as to restore the signal before the channel. This receiver is called Zero Forcing as it brings down the ISI to zero [5]. The frequency response of channel is assumed to be $F(f)$ and $C(f)$ for the zero forcing equalizer, then this equalizer is constructed such that $C(f) = 1 / F(f)$. Thus this combination of channel and equalizer gives a flat frequency response and linear phase.

The received signal can be represented by using the linear model as:

$$y = Hx + n \tag{1.7}$$

A 2x2 MIMO channel can be represented in matrix notation as follows:

$$\begin{pmatrix} y_1 \\ y_2 \end{pmatrix} = \begin{pmatrix} h_{1,1} & h_{1,2} \\ h_{2,1} & h_{2,2} \end{pmatrix} \begin{pmatrix} x_1 \\ x_2 \end{pmatrix} + \begin{pmatrix} n_1 \\ n_2 \end{pmatrix} \tag{1.8}$$

The signal received on the first receive antenna can be expressed as:

$$y_1 = h_{1,1} x_1 + h_{1,2} x_2 + n_1 = [h_{1,1} \quad h_{1,2}] \begin{bmatrix} x_1 \\ x_2 \end{bmatrix} + n_1 \tag{1.9}$$

The signal received on the second receive antenna can be expressed as:

$$y_2 = h_{2,1} x_1 + h_{2,2} x_2 + n_2 = [h_{2,1} \quad h_{2,2}] \begin{bmatrix} x_1 \\ x_2 \end{bmatrix} + n_2 \tag{1.10}$$

where
x_1 and y_1 is the transmitted and received symbol on the first antenna,
x_2 and y_2 is the transmitted and received symbol on the second antenna,
$h_{1,1}$ is the channel from 1st transmit antenna to the 1st receive antenna,
$h_{1,2}$ is the channel from 2nd transmit antenna to the 1st receive antenna,
$h_{2,1}$ is the channel from 1st transmit antenna to the 2nd receive antenna,
$h_{2,2}$ is the channel from 2nd transmit antenna to the 2nd receive antenna,
and n_1, n_2 are the noise on 1st and 2nd receive antennas.

VII.   SIMULATED RESULTS

In this section, the BER analysis of MIMO system structure is done for M-PSK Modulation techniques over AWGN and Rician fading channels using Space-Time Block Coding (STBC) structure. The BER analysis of MIMO system is done for M-PSK modulation for different values of M. Here the value of M selected can be 32, 64, 128, 256, 512 and 1024 over both the fading channels.

(A)M-PSK over AWGN channel

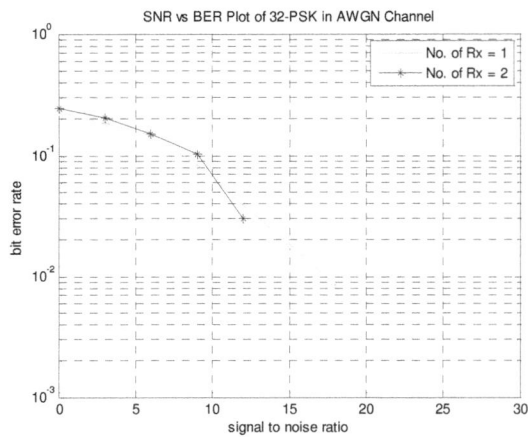

(a) 32-PSK

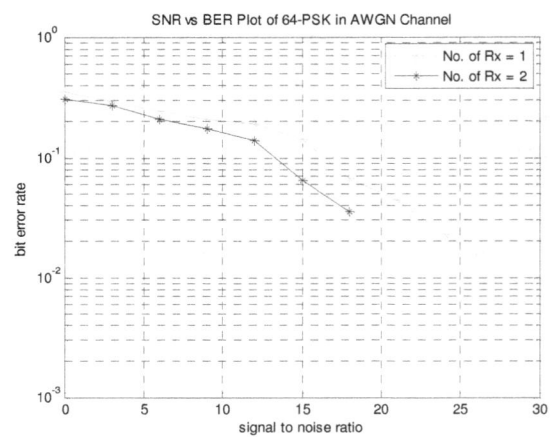

(b) 64-PSK

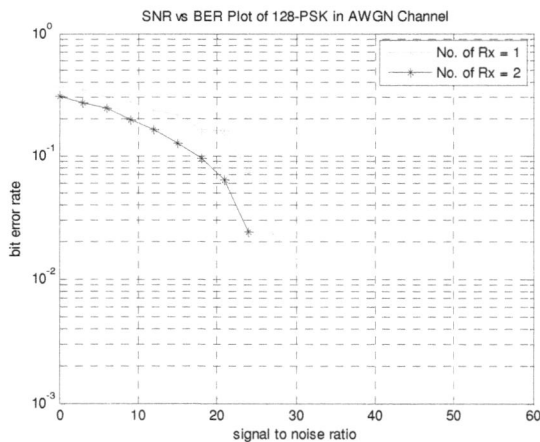

(c) 128-PSK

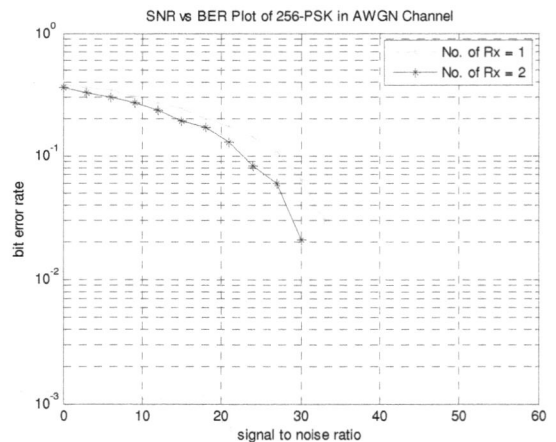

(d) 256-PSK

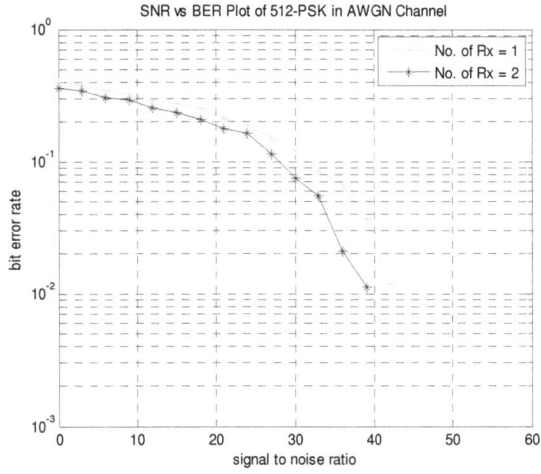

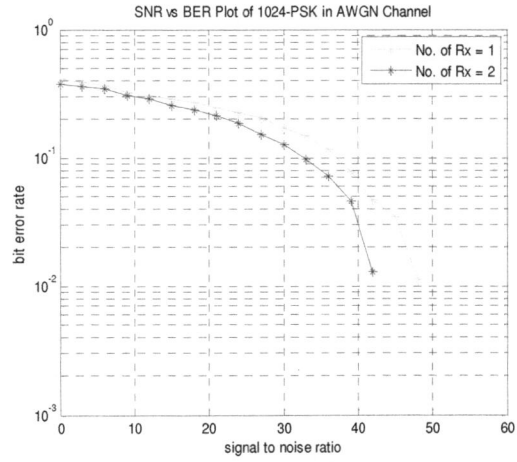

(e) 256-PSK   **(f) 1024-PSK**

Fig. 1.4: SNR vs BER plots using M-PSK over AWGN channel for different values of M

In Fig. 1.4 (a) – (f), the SNR vs BER plots using M-PSK over AWGN channel for MIMO system are presented for different values of 'M' employing different antenna configurations. It can be concluded from the graphs that with the increase in number of receiving antennas, the BER keeps on decreasing due to space diversity in MIMO and thus the system proposed over here provide better BER performance in comparison to the other antenna configurations.

(B) M-PSK over Rician channel

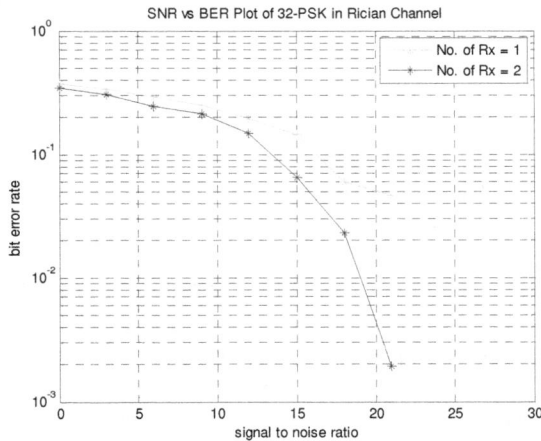

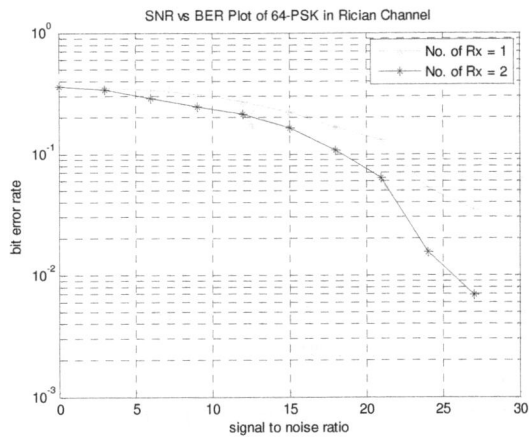

(a) 32-PSK   **(b) 64-PSK**

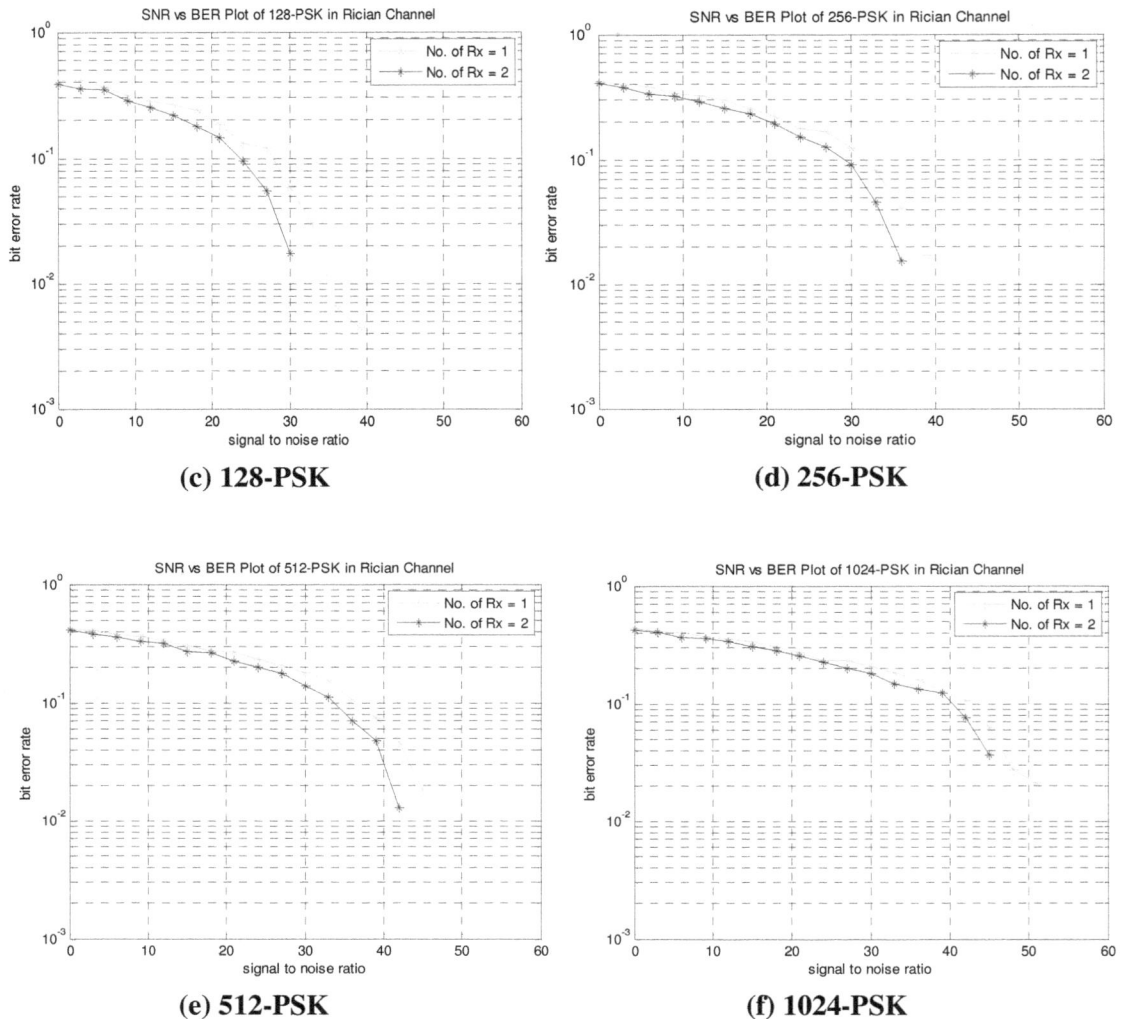

(c) 128-PSK

(d) 256-PSK

(e) 512-PSK

(f) 1024-PSK

Fig. 1.5: SNR vs BER plots using M-PSK over Rician channel for different values of M

The SNR vs BER plots using M-PSK over Rician channel for MIMO system are presented for different values of 'M' employing different antenna configurations are presented in Fig. 1.5 (a) – (f). From the graphs, it can be seen that if there is increase in the number of receiving antennas in MIMO system then the BER keeps on decreasing due to space diversity. Thus this system provides better BER performance as compared to the other antenna configurations.

VIII. CONCLUSION

In this paper, SNR vs. BER plots for M-PSK over AWGN and Rician fading channels for MIMO system employing different antenna configurations are presented. It can be concluded that in MIMO system, the BER keeps on decreasing due to space diversity as we goes on increasing the number of receiving antennas and the proposed system provide better BER performance. But BER is greater in Rician channel as compared to that of AWGN channel. Also as we goes on increasing the value of M for M-PSK i.e the no. of constellation points in the constellation diagram, the BER is also increasing. This increase in BER is due to the fact that as increase the size of constellation diagram the spacing in between different constellation point will keep on decreasing, which results in decreasing the width of decision region for each constellation point which in turn makes the detection of the signal corresponding to the constellation point much tougher. Due to this fact the BER is increasing as we goes on increasing the number of points in constellation diagram.

IX. REFERENCES

[1] P. Sanghoi & L. Kansal, "Analysis of WIMAX Physical layer Using Spatial Diversity", International Journal of Computer Application, Vol. 44, Issue 5, 2012.

[2] L. Kansal, A. Kansal & K. Singh, "BER Analysis of MIMO-OFDM Sytem Using OSTBC Code Structure for M-PSK under Different fading Channels", International Journal of Scientific & Engineering Research, Vol. 2, Issue 11, 2011.

[3] P. Sanghoi & L. Kansal, "Analysis of WIMAX Physical layer Using Spatial Diversity under different Fading Channels", International Journal of Computer Application, Vol. 44, Issue 20, 2012.

[4] S. Alamouti, "A simple transmit diversity technique for wireless communications", IEEE Journal on Selected Areas of Communication, Vol. 16, Issue 8, pp. 1451–1458, 1998.

[5] V. Tarokh, H. Jafarkhani & A. R. Calderbank, "Space–time block codes from orthogonal designs", IEEE Transactions on Information Theory, Vol. 45, Issue 5, pp. 1456–1467, 1999.

[6] A. I. Sulyman, "Performance of MIMO Systems With Antenna Selection Over Nonlinear Fading Channels", IEEE Journal of Selected Topics in Signal Processing, Vol. 2, Issue 2, pp. 159-170, 2008.

[7] S. G. Kim, D. Yoon, Z. Xu & S. K. Park, "Performance Analysis of the MIMO Zero-Forcing Receiver over Continuous Flat Fading Channels", IEEE Journal of Selected Areas in Communications, Vol. 20, Issue 7, pp. 324 – 327, 2009.

[8] C. Wang, "On the Performance of the MIMO Zero-Forcing Receiver in the Presence of Channel Estimation Error", IEEE Transactions on Wireless Communications, Vol. 6, Issue 3, pp. 805 – 810, 2007.

[9] X. Zhang, Z. Lv & W. Wang, "Performance Analysis of Multiuser Diversity in MIMO Systems with Antenna Selection", IEEE Transactions on Wireless Communications, Vol. 7, Issue 1, pp. 15-21, 2008.

[10] A. Lozano & N. Jindal, "Transmit Diversity vs. Spatial Multiplexing in Modern MIMO Systems", IEEE Transactions on Wireless Communications, Vol. 9, Issue 1, pp. 186-197, 2010.

[11] P. W. Wolniansky, G. J. Foschini, G. D. Golden & R. A. Valenzuela, "V-Blast: An architecture for realizing very high data rates over the rich-scattering channel", International Symposium on Signals, Systems and Electronics, pp. 295–300, 1998.

[12] R. S. Blum, Y. Li, J. H. Winters & Q. Yan, "Improved Space–Time Coding for MIMO-OFDM Wireless Communications", IEEE Transaction on Communications, Vol. 49, Issue 11, pp. 1873-1878, 2001.

[13] C. Chen, "Performance Analysis of Scheduling in Multiuser MIMO Systems with Zero-Forcing Receivers", IEEE Journal of Selected Areas in Communications, Vol. 25, Issue 7, pp. 1435–1445, 2007.

[14] N. S. Kumar, G. J. Foschini, G. D. Golden & R. A. Valenzuela, "Bit Error Rate Performance Analysis of ZF, ML and MMSE Equalizers for MIMO Wireless Communication Receiver", European Journal of Scientific Research, Vol. 59, Issue 4, pp. 522–532, 2011.

4

IMPROVE PERFORMANCE OF TCP NEW RENO OVER MOBILE AD-HOC NETWORK USING ABRA

Dhananjay Bisen[1] and Sanjeev Sharma[2]

[1]M.Tech, School Of Information Technology, RGPV, BHOPAL, INDIA
[1]bisen.it2007@gmail.com
[2]Reader & Head, School Of Information Technology, RGPV, BHOPAL, INDIA
[2]sanjeev@rgtu.net

ABSTRACT

In a mobile ad hoc network, temporary link failures and route changes occur frequently. With the assumption that all packet losses are due to congestion, TCP performs poorly in such an environment. There are many versions of TCP which modified time to time as per need. In this paper modifications introduced on TCP New Reno over mobile ad-hoc networks using calculation of New Retransmission Time out (RTO), to improve performance in term of congestion control. To calculate New RTO, adaptive backoff response approach (ABRA) in TCP New Reno was applied which suggest ABRA New Reno. It utilizes an ABRA by which congestion window and slow start threshold values were decreased whenever an acknowledgement is received and new backoff value calculate from smoothed round trip time. Evaluation based on comparative study of ABRA New Reno with other TCP Variants like New Reno and Reno was done using realistic parameters like TCP Packet Received, Packet Drop, Packets Retransmitted, Throughput, and Packet Delivery Ratio calculated by varying attributes of Node Speed, Number of Nodes and Pause Time. Implementation and simulations were performed in QualNet 4.0 simulator.

KEYWORDS

Mobile ad hoc network, RTO, TCP New Reno, TCP Tahoe, ABRA New Reno, Congestion control, TCP Timer

1. INTRODUCTION

Mobile ad hoc networks (MANETs) [1] are collections of mobile nodes, dynamically forming a temporary network without centralized administration. These nodes can be arbitrarily located and are free to move randomly at any given time, thus allowing network topology and interconnections between nodes to change rapidly and unpredictably. There has been significant research activity over the past 10 year into performance of such networks with the view to develop more efficient and robust TCP variants. Transmission control protocol [2] provides reliability, end-to-end congestion control mechanism, byte stream transport mechanism, flow control, and congestion control. Comparing to wire networks, there are many different characteristics in wireless environments, which makes TCP congestion control mechanism is not directly suitable for wireless networks and many improved TCP congestion control mechanisms [3] have been presented. However, TCP in its present form is not well suited for mobile ad hoc networks. In addition to all links being wireless, frequent route failures due to mobility can cause serious problems to TCP as well. Route failures can cause packet drops at the intermediate nodes, which will be misinterpreted as congestion loss.

When a route failure occurs for a period of time greater than retransmission timer value [4] [8], TCP understand this as congestion which means decreasing both the congestion window (CWND) and slow start threshold (SSThr). Subsequently it retransmits the first unacknowledged packet and executes back-off by doubling the value of retransmission timer. Under multiple successive back-offs, the value of the retransmission timer is too long, However during the long retransmission period, the route may come back but TCP will not try to retransmit the first unacknowledged packet until the retransmission timer expires. The ideal solution for the route failure problem is to freeze its state as soon as the route breaks and resume as soon as a new route is found. Most of research are concentrating on congestion, corruption control and improve retransmission time out condition, improving performance metrics and security threats of protocol. Hence this thrust area of mobile ad-hoc network become the choice of interest for us. This paper is organized in such a manner that section 2 describes Congestion Control Algorithm followed by TCP New Reno, Adaptive Backoff Response Approach and new retransmission time out in section 3, 4 and 5 respectively. In section 6, simulation environment and methods is described with simulation results and analysis of the performance in 7.

2. CONGESTION CONTROL ALGORITHM

TCP is a transport layer protocol used by applications that require guaranteed delivery and reliable transmission of packets. It is a sliding window protocol that provides handling for both timeouts and retransmissions. Another function is Congestion control [6] [7]. It is a method used for monitoring the process of regulating the total amount of data entering the network so as to keep traffic levels at an acceptable value, whereas flow control controls the per-flow traffic such that the receiver capacity is not exceed. Congestion control mostly applies to packet-switching network. A wide variety of approaches have been proposed, however the "objective is to maintain the number of packets within the network below the level at which performance falls off dramatically."

The TCP sender starts the session with a congestion window value of one maximum segment size (MSS) it is a parameter of the TCP protocol that specifies the largest amount of data, specified in bytes, that a computer can receive in a single, unfregmented   piece. The window size determines the number of bytes of data that can be sent before an acknowledgement from the receiver is necessary. This doubling of the congestion window with every successful acknowledgment of all the segments in the current congestion window is called slow-start or exponential start show in figure 1 and it continues until the congestion window reaches the slow-start threshold.

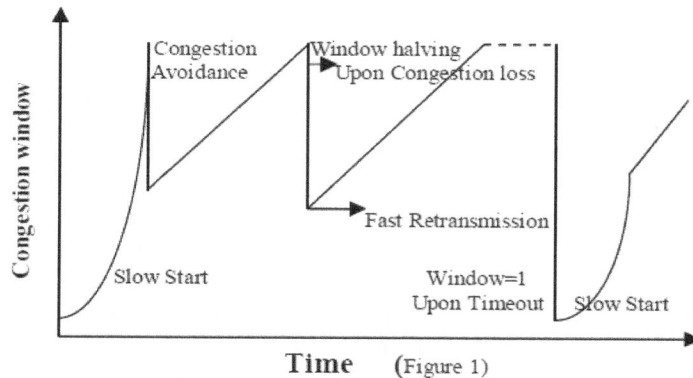

Time   (Figure 1)

Once it reaches the slow-start threshold, it grows linearly, adding one MSS to the congestion window on every ACK received. This linear growth, which continues until the congestion window reaches the receiver window, is called congestion avoidance, [5] show in figure 1 as it tries to avoid increasing the congestion window exponentially, which will surely worsen the congestion in the network.

TCP updates the RTO period with the current round-trip delay calculated on the arrival of every ACK packet. If the ACK packet does not arrive within the RTO period, then it assumes that the packet is lost due to the congestion in the network and if TCP sender receives three consecutive duplicate ACKs (DUPACKs). Upon reception of three DUPACKs, the TCP sender retransmits the oldest unacknowledged segment. This is called the fast retransmit scheme show in figure 1. The TCP sender does the following during congestion: Reduces the slow-start threshold to half the current congestion window, Resets the congestion window size to one MSS,

Activates the slow-start algorithm, and Resets the RTO with an exponential back-off value which doubles with every subsequent retransmission.

A variety of TCP approaches have been proposed like TCP Tahoe [6], TCP Reno [9], TCP with selective ACK (SACK) [9]. There are used various congestion control algorithms slow start threshold, congestion avoidance, congestion detection, fast recovery, fast retransmission. This is for detect losses and congestion. That is, TCP uses timeout and duplicate ACKs to detect loss, and changes in round-trip times to detect congestion.

3. TCP NEW RENO

The experimental version of TCP Reno is known as TCP New Reno [5]. It is slightly different than TCP Reno in fast recovery algorithm. New Reno is more competent than Reno when multiple packets losses occur. New Reno and Reno both correspond to go through fast retransmit when multiple duplicate packets received, but it does not come out from fast recovery phase until all outstanding data was not acknowledged . It implies that in New Reno, partial ACK do not take TCP out of fast recovery but they are treated as an indicator that the packet in the sequence space has been lost, and should be retransmitted. Therefore, when multiple packets are lost from a single window of data, at this time New Reno can improve without retransmission time out. The retransmitting rate is one packet loss per round trip time until all of the lost packets from that window have been transmitted. It exist in fast recovery till all the data is injected into network, and still waiting for an acknowledgement that fast recovery was initiated.

The critical issue in TCP New Reno [12] is that it is capable of handling multiple packet losses in a single window. It is limited to detecting and resending only one packet loss per round - trip-time. This insufficiency becomes more distinct as the delay-bandwidth becomes greater. However, still there are situations when stalls can occur if packets are lost in successive windows, like all of the previous versions of TCP New Reno which infer that all lost packets are due to congestion and it may therefore unnecessarily cut the congestion window size when errors occur. There are [12] some steps of congestion control for New Reno transmission control protocol.

Step 1: Initially

    0<cwnd<= min (4*mss, max (2*mss, 4380 bytes))

    SS_threshold = max (cwnd/2, 2*MSS)

Step 2: Slow Start Algorithm (Exponential Increases)

    If (receive acks && cwnd< ss_threshold)

    cwnd = cwnd+1;

Step 3: Congestion Avoidance Algorithm (Additive Increase)

    If (receive ACKs) {

    If (cwnd > ss_threshold)

      cwnd = cwnd + segsize * segsize / cwnd;

    Else

      cwnd = cwnd + 1;   }

Step 4: Congestion Detection Algorithm (Multiplicative Decrease): Fast Retransmission and Fast Recovery

    If (congestion) {

    If (Receive same Acks 3 time or retransmission time out) {

    SS_threshold = cwnd/2;

    If (Retransmission time out) {

        cwnd = initial;

    Exit and call Slow Start step;

    Else /* Receive same Acks 3 time*/

        cwnd = SS_threshold;

    Exit and call congestion avoidance step;   }}}

4. ADAPTIVE BACKOFF RESPONSE APPROACH (ABRA)

In the event of a retransmit timeout, TCP retransmits the oldest unacknowledged packet and doubles the retransmit timeout interval (RTO). This process is repeated until an ACK for the retransmitted packet has been received. So retransmission timeout interval may be very long although the route may have been re-established some time ago. This leads to a wasted time. The wasted time can be used to send packets. We try to make use of this wasted time by making the retransmission timeout interval depends on a smoothed round trip time (SRTT), which is a weighted average of measured retransmitted timeout. The ABRA [10] depends on saving the values of congestion window, slow start threshold and smoothed round trip time, when the retransmission timer expires. Here instead of multiplying RTO interval by two each time, we multiply it by a value called backoff $_{new}$ between one and two depending on the last_srtt, which is a weighted average of last measured retransmitted timeout.. The backoff $_{new}$ and the new RTO value (RTO$_{new}$) are computed as follows:

$$Backoff_{new} = 1 + \frac{(last\_srtt - min\_srtt)}{(max\_srtt - min\_srtt)}$$
$$RTO_{new} = Backoff_{new} * RTO_{current}$$

where RTO $_{current}$ is the current RTO value, min_srtt is the minimum smoothed round trip time seen so far, and max_srtt is the maximum smoothed round trip time seen so far. We initialize [10] min_srtt and max_srtt with the values 0.1 and 0.6 seconds respectively choosing initialization values of these two variables.

5. NEW RETRANSMISSION TIMEOUT CALCULATION

There are steps of calculation of new retransmitted time out using Adaptive Backoff Response approach for New Reno transmission control protocol.

Step 1: Calculate Round Trip Time (RTT):- the measured RTT for a segment is the time required for the segment to reach the destination and be acknowledged.
Step 2: Calculate Smoothed Round Trip Time (SRTT) [8]: which is a weighted average of RTT $_{measured}$ and the previous SRTT as shown below:
For first measurement
$SRTT_{previous} = RTT_{measured}$
From second measurement the calculus becomes
$SRTT = (1-a)\,SRTT_{previous} + a\,.RTT_{measured}$
   $a = 1/8$
Step 3: Calculate RTT (Deviation): RTTD [8]
For first measurement
$RTTD_{previous} = RTT_{measured}/2$
From second measurement the calculus becomes
$RTTD = (1-b)\,RTTD_{previous} + b.\,(SRTT - RTT_{measured})$
   $b = 1/4$
Step 4: Calculate Retransmission Time out (RTO): The value of RTO is based on the smoothed round trip time and its deviation.
For first measurement
$RTO_{current} = SRTT + 4.RTTD$
Calculate New RTO with using SRTT; apply Adaptive back off response approach.
Step 5:(Proposed Step) for second measurement, first calculate New Back-off value and multiply that Value to RTO (previous RTO value).

$$Backoff_{new} = 1 + \frac{(last\_srtt - min\_srtt)}{(max\_srtt - min\_srtt)}$$
$$RTO_{new} = Backoff_{new} * RTO_{current}$$

Constant a and b in step 2 and step 3 have been selected on basis of the literature where a=1/8 and b=1/4 is mentioned in reference [8] as well as value of a and b are implementation dependent but under normal circumstances they are set to be 1/8 and 1/4 respectively.

6. SIMULATION ENVIRONMENT

All the simulation work is perform in QualNet wireless network simulator version 4.0 [13]. Initially number of nodes are 50, Simulation time was taken 200 seconds and seed as 1. All the scenarios have been designed in 1500m x 1500m area. Mobility model used is Random Way Point [14] (RWP). The mobility model is designed to describe the movement pattern of mobile users, and how their location, velocity and acceleration change over time. In random-based mobility simulation models, the mobile nodes move randomly and freely without restrictions. To be more specific, the destination, speed and direction are all chosen randomly and independently of other nodes. For simulation, environmental surrounding selected are speed, Pause time and no. of mobile nodes. Speed and pause time are between the range of 5-30 m/s and sec. respectively while no. of nodes are between the range of 10-50. "Pause time is a time in which all nodes in network are motionless but transmission in continued". All the simulation work were carried out using TCP variants (Reno, New Reno, ABRA New Reno) with AODV routing protocol .Network traffic is provided by File Transfer Protocol (FTP) application. File Transfer Protocol represents the File Transfer Protocol server and client.

7. RESULTS AND ANALYSIS

The simulation for ABRA New Reno is based on simulation time, number of node, area of network, pause time, routing protocols, and speed of node. In experimental methodologies variation in one of the parameter including no. of node, speed of node and the pause time was done each time and their effect on performance of different TCP protocols were determined while rest of all other parameters like simulation time, area of network, pause time, routing protocols, and speed of node kept constant. Effects of simulation studies on performance of ABRA New Reno and other TCP protocols under experimental conditions mentioned above were represented graphically.

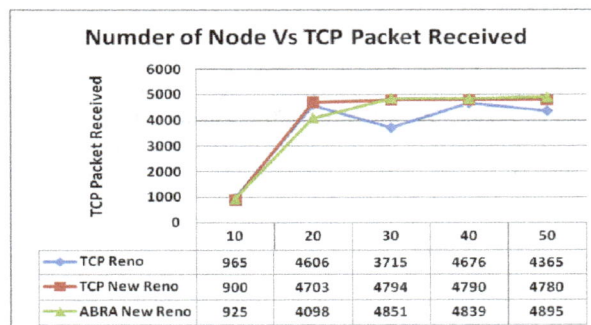

Number of Node Vs TCP Packet Received

	10	20	30	40	50
TCP Reno	965	4606	3715	4676	4365
TCP New Reno	900	4703	4794	4790	4780
ABRA New Reno	925	4098	4851	4839	4895

Figure 2

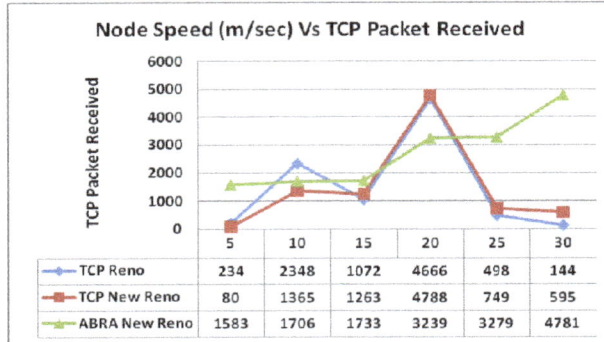

	5	10	15	20	25	30
TCP Reno	234	2348	1072	4666	498	144
TCP New Reno	80	1365	1263	4788	749	595
ABRA New Reno	1583	1706	1733	3239	3279	4781

Figure 3

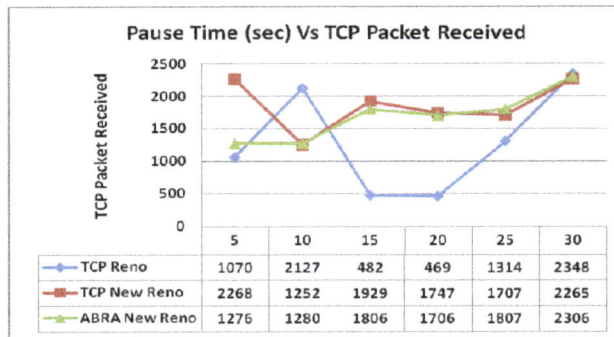

	5	10	15	20	25	30
TCP Reno	1070	2127	482	469	1314	2348
TCP New Reno	2268	1252	1929	1747	1707	2265
ABRA New Reno	1276	1280	1806	1706	1807	2306

Figure 4

Simulation results in fig. 2-4, it is observed that under less no. of nodes, node speed and pause time sender does not find proper path, hence, receiver is far from source and no. of mobile nodes are least between sender and receiver, hence, time out will crop up, and therefore minimum packet received. But upon increments in variants like no. of nodes, node speed and Pause time of the network then sender will receive different paths between senders and receiver easily, hence maximum packets were received.

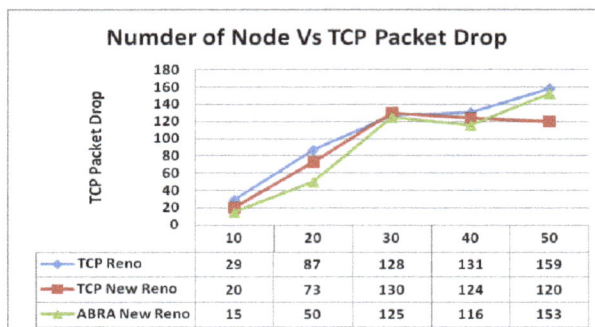

	10	20	30	40	50
TCP Reno	29	87	128	131	159
TCP New Reno	20	73	130	124	120
ABRA New Reno	15	50	125	116	153

Figure 5

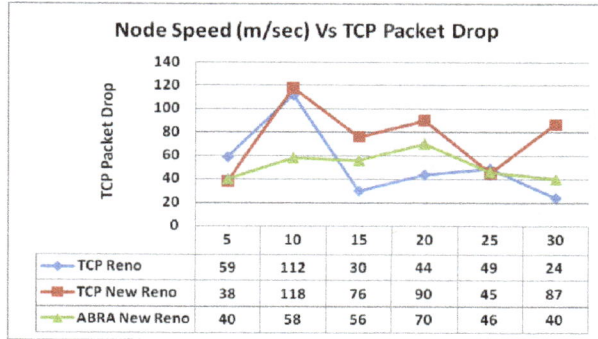

Node Speed (m/sec) Vs TCP Packet Drop

	5	10	15	20	25	30
TCP Reno	59	112	30	44	49	24
TCP New Reno	38	118	76	90	45	87
ABRA New Reno	40	58	56	70	46	40

Figure 6

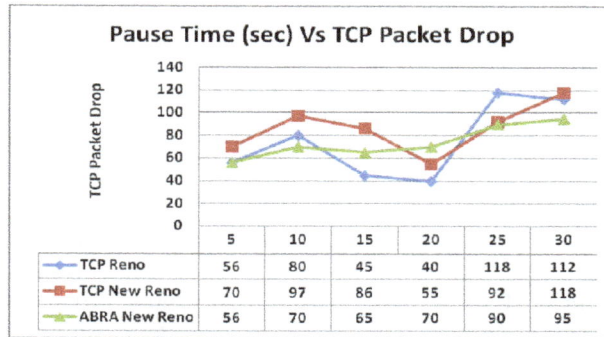

Pause Time (sec) Vs TCP Packet Drop

	5	10	15	20	25	30
TCP Reno	56	80	45	40	118	112
TCP New Reno	70	97	86	55	92	118
ABRA New Reno	56	70	65	70	90	95

Figure 7

Simulation results in fig. 5, 6 & 7 shows that initially Packets drops are less because of less congestion, less timeout waiting time, low route failure, but upon increments in variants like no. of nodes, node speed and Pause time of the network, packet drop rate is also increases gradually because of Route Failure, high Congestion, destination unreachable, time slice expired, frequently change route and bandwidth, improper routing path and node mobility.

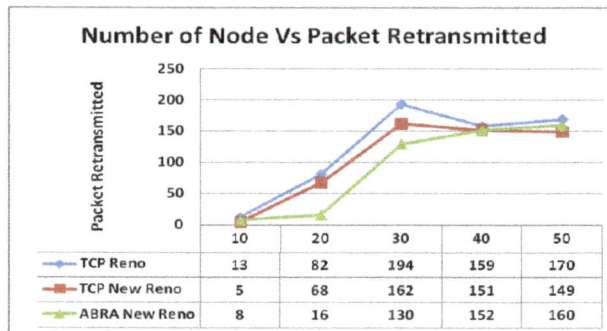

Number of Node Vs Packet Retransmitted

	10	20	30	40	50
TCP Reno	13	82	194	159	170
TCP New Reno	5	68	162	151	149
ABRA New Reno	8	16	130	152	160

Figure 8

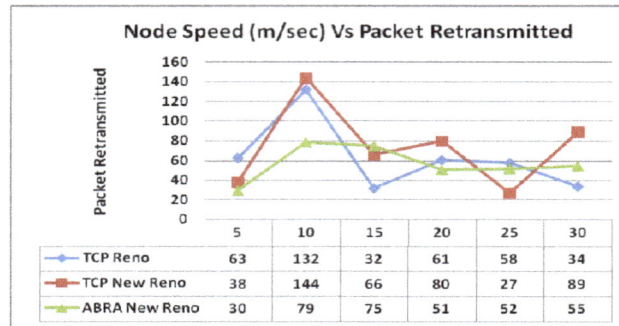

Figure 9

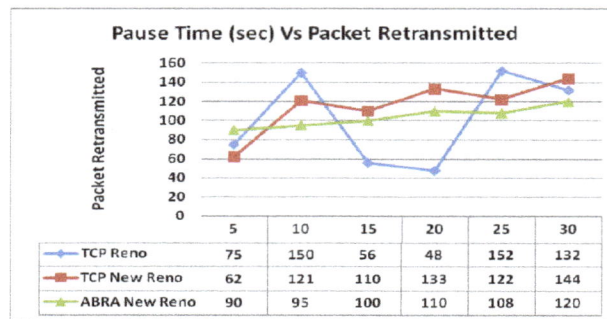

Figure 10

Simulation results in fig. 8-10 shows packet retransmitted upon different variants like variation in number of node speed of node and pause time in network. When number of nodes increases from 10 to 40 in network than packet retransmission is also increases gradually, but Packets retransmission rate vary under increase node mobility and pause time due to increments in packet drop rate, acknowledgment loss, packet delay and also by time out condition.

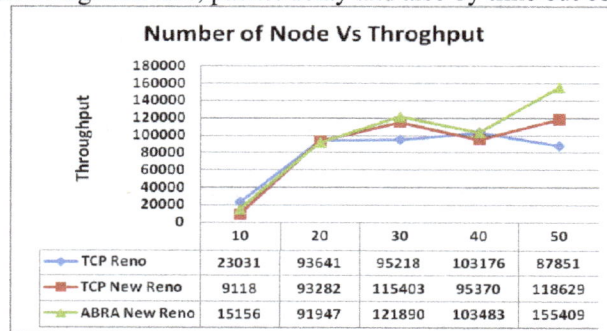

Figure 11

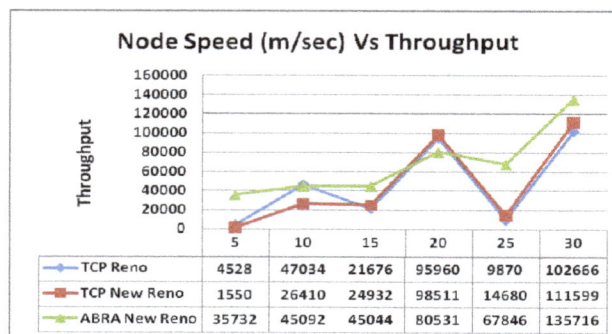

Figure 12

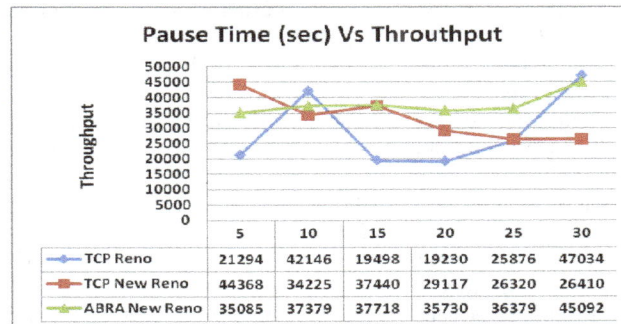

Figure 13

Simulation results in fig. 11-13 shows that throughput of different variants with variation in number of nodes, speed of node and pause time in network. It is observed that throughput of ABRA New Reno TCP is vary according to variation in number of node, speed of node and Pause time. Throughput depends on receive bytes, packet loss and drop acknowledgement. These parameters are inconsistent and greater sometime, because of wrong estimation of bandwidth and delay.

8. CONCLUSION

An improve TCP New Reno (ABRA New Reno) is proposed for mobile ad-hoc networks using calculations of New Retransmission Time out, to improve performance in terms of congestion control and implemented in a Mobile Ad-hoc Network under QualNet 4.0 simulator. To calculate New RTO, adaptive backoff response approach was applied in TCP New Reno which suggests ABRA New Reno. Extensive simulation studies were taken to compare its performance with standards TCP Reno and TCP New Reno over Ad-hoc Mobile Network. After implementation ABRA in new Reno it is analysed under varying conditions of node speed, Pause Time and number of node. Simultaneously efficiency of ABRA New Reno on various performance metrics was measured, including data packet received, packet drop, packets retransmitted, throughput.

From results it is observed that, ABRA New Reno performs well under the varying conditions of high density node, high node speed and pause time, because of proper utilization of time, optimal paths between nodes, optimal bandwidth exploitation and less packet delay. Simultaneously throughput was found to be most favourable, however it may be vary according to traffic conditions and congestion.

9. FUTURE WORK

Our experimental results suggest that in forthcoming efforts, simulation of ABRA New Reno TCP by varying other parameter and performance in large and realistic scenario will have great potential. Furthermore simulations are to be done in future to explore the effect of ABRA New Reno TCP on other TCP Variants and ad hoc routing protocols. In order to accurately simulate the realistic congested network environment, there is a need to experiment with multiple TCP flows.

10. REFERENCES

[1] Imrich Chlamtac, Marco Conti, Jennifer J.-N. Liu .Mobile ad hoc networking: imperatives and challenges, School of Engineering, University of Texas at Dallas, Dallas, TX, USA ,b Istituto IIT, Consiglio Nazionale delle Ricerche, Pisa, Italy ,c Department of Computer Science, University of Texas at Dallas, Dallas, TX, USA. Ad Hoc Networks 1 (2003)

[2] Hanbali, Eitan Altman, And P. Nain, "A Survey of TCP over Ad Hoc Networks," IEEE communications Surveys Tutorials, vol 7, Third Quarter 2005, no. 3.

[3] Liu and S. Singh. "ATCP: TCP for mobile ad hoc networks," IEEE Journal on Selected Areas in Communications, vol. 19, no. 7, July 2001, pp. 1300-1315.

[4] Haifa Touati, Ilhem Lengliz, Farouk Kamoun TCP Adaptive RTO to improve TCP performance in mobile ad hoc networks. June 12-15, 2007.

[5] S. Floyd, the New Reno Modification to TCP's Fast Recovery Algorithm, RFC 3782, April 2004.

[6] W. Stevens. TCP slow start, congestion avoidance, fast retransmit, and fast recovery algorithms. IETF RFC 2001, January 1997.

[7] Fu Lin DengYi Zhang, Wenbin Hu "An Improved TCP Congestion Control Mechanism Based on Double-Windows for Wireless Network", 2008.

[8] V.Paxson and M. Allman, "Computing TCP's Retransmission Timer," RFC 2988, Standards Track, NASA GRC/BBN, November 2000.

[9] Jinwen Zhu, Tianrui Bai , Performance of Tahoe, Reno, and SACK TCP at Different Scenarios, 27-30 Nov. 2006, On page(s): 1-4.

[10] Tamer F. Ghanem, Wail S. Elkilani, Mohiy M. Hadhoud Improving TCP Performance over Mobile Ad Hoc Networks Using an Adaptive Backoff Response Approach, 978-1-4244-3778-8/09/2009 IEEE.

[11] F. Wang and Y. Zhang, "Improving TCP Performance over Mobile Ad Hoc Networks with Out-of-order Detection and Response," Proc. ACM MOBIH-OC, Lausanne, Switzerland, June 2002, pp. 217-25.

[12] Qureshi, M. Othman, Member, IEEE, and N. A. W. Hamid Progress in Various TCP Variants (February 2009).

[13] Scalable Network Technology, "QualNet4.0 simulator" tutorial and QualNet Forum,http://www.scalable-networks.com/forums/

[14] Guolong Lin, Guevara Noubir, and Rajmohan Rajaraman, College of Computer & Information Science. Mobility Models for Ad hoc Network Simulation 0-7803-8356-7/04, 2004 IEEE.

5

MADSN: Mobile Agent Based Detection of Selfish Node in MANET

Debdutta Barman Roy[1] and Rituparna Chaki[2]

[1]Department of Information Technology, Calcutta Institute of Eng. & Mgmt , Kolkata,
India
barmanroy.debdutta@gmail.com
[2]Department of Computer Science and Eng., West Bengal University of Technology,
Kolkata, India
rituchaki@gmail.com

ABSTRACT

Mobile Adhoc Network (MANET) is highly vulnerable to attacks due to the open medium dynamically changing network topology, co-operative algorithm, lack of centralized monitoring and management point. The fact that security is a critical problem when implementing mobile ad hoc networks (MANETs) is widely acknowledged. One of the different kinds of misbehavior a node may exhibit is selfishness. Routing protocol plays a crucial role for effective communication between mobile nodes and operates on the basic assumption that nodes are fully cooperative. Because of open structure and limited battery-based energy some nodes (i.e. selfish or malicious) may not cooperate correctly. There can be two types of selfish attacks —selfish node attack (saving own resources) and sleep deprivation (exhaust others' resources. In this paper, we propose a new Intrusion Detection System (IDS) based on Mobile Agents. The approach uses a set of Mobile Agent (MA) that can move from one node to another node within a network. This as a whole reduces network bandwidth consumption by moving the computation for data analysis to the location of the intrusion. Besides, it has been established that the proposed method also decreases the computation overhead in each node in the network.

KEYWORDS

MANET, Mobile Agent, Selfish Node, IDS

1. INTRODUCTION

A Mobile Ad hoc Network (MANET) is an autonomous system, in which mobile hosts connected by wireless links are free to move randomly and often act as a router sometimes [4,5].Therefore the limited wireless transmission range of each node gets executed by multi-hop packet forwarding. That is here nodes within each other's radio range communicate directly via wireless links while those are far apart uses other nodes as relays. This kind of network is well suited for the mission critical applications such as emergency relief, military operations where no pre-deployed infrastructure exists for communication. Due to the lack of authorization facilities, volatile network topology it is hard to detect malicious nodes [4, 5], MANETs are highly vulnerable to attacks. Finally, in

 A MANET nodes might be battery-powered and might have very limited resources, which may make the use of heavy-weight security solutions undesirable [7, 8, 9, 10 and 11].
Many different types of attacks have been identified. This paper deals with the Denial of service attack (DoS) by a selfish node; this is the most common form of attack which decreases the network performance.

A selfish node does not intend to directly damage other nodes, but is unwilling to spend battery life, CPU cycles, or available network bandwidth to forward packets not of direct interest to it, even though it expects others to forward packets on its behalf. The reason behind this is 'saving one's own resource' by saving of battery power, CPU cycles or protecting wireless bandwidth in certain direction. A selfish node wants to preserve own resources while using the services of others and consuming their resources. Detecting routes and forwarding packets consumes local CPU time, memory, network-bandwidth, and last but not least energy. Therefore there is a strong motivation for a node to deny packet forwarding to others, while at the same time using their services to deliver own data.

According to the attacking technique the selfish node can be defined in three different ways [1]

SN1: These nodes take participation in the route discovery and route maintenance phases but refuses to forward data packets to save its resources.

SN2: These nodes neither participate in the route discovery phase nor in data-forwarding phase. Instead they use their resource only for transmissions of their own packets.

SN3: These nodes behave properly if its energy level lies between full energy-level E and certain threshold T1. They behave like node of type SN2 if energy level lies between threshold T1 and another threshold T2 and if energy level falls below T2, they behave like node of type SN1

One immediate effect of node misbehaviors and failures in wireless ad hoc networks is the node isolation problem due to the fact that communications between nodes are completely dependent on routing and forwarding packets. In turn, the presence of selfish node is a direct cause for node isolation and network partitioning, which further affects network survivability. Traditionally, node isolation refers to the phenomenon in which nodes have no (active) neighbors; however, we will show that due to the presence of selfish node, a node can be isolated even if active neighbors are available [2].

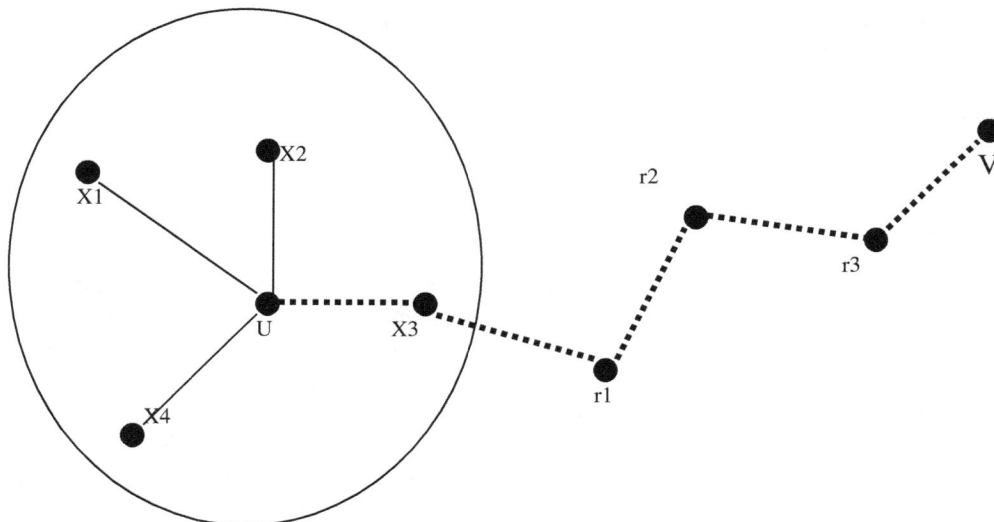

Fig 1: Node isolation due to selfish neighbors

In Figure.1, suppose node x3 is a selfish node. When node u initiates a route discovery to another node v, the selfish neighbors x3 may be reluctant to broadcast the route request from u. In this case, x3 behaves like a failed node. It is also possible for x3 to forward control packets; however, the situation could be worse since u may select x3 as the next hop and send data to it. Consequently, x3 may discard all data to be forwarded via it, and then communications between u and v cannot proceed. When all neighbors of u are selfish, u is unable to establish any communications with other nodes at a distance of more than one-hop away. In this case, we say that a node is isolated by its selfish neighbors. Note that selfish nodes can still communicate with other nodes (via their cooperative neighbors), which is different from failed nodes.

2. RELATED WORK

Several methods proposed to defend these attacks have been studied. These can be classified into three types: reputation based scheme, credit based approach and game theoretic approach [1] [3] [6]

2.1 Reputation Based scheme

In a reputation based scheme [1] watchdog and path rater approach the IDS overhear neighbors' packet transmission promiscuously and notify misbehavior to the source node by sending a message. The source node collects the notifications and rates every other node to avoid unreliable nodes in finding a path. Though the scheme is easier to implement but it depends only on promiscuous listening that may results false identification.

CONFIDANT (Cooperation of Nodes, Fairness in Dynamic Ad-hoc Networks), in this scheme the IDS performs task in a distributed ways the monitor node promiscuously observes route protocol behavior as well as packet transmission of neighbor node. The Trust manager sends ALARM messages on detection of misbehavior. The Reputation system: maintains a rating list and a blacklist for other nodes. The Path manager ranks paths according to the reputation of nodes along each path. This scheme uses both direct and indirect observations from other nodes. In this scheme the adversary nodes are black listed but not removed from the network. As the detection depends on the other nodes that reduces the reliability of the IDS because any one of the above mentioned nodes may provide false result that may blacklisted a nonadversary node.

CORE (Collaborative Reputation) approach, here the source node observes usual packet transmission and the task specific behavior of neighbor nodes and rate the node by using the positive reports from other nodes. The malicious node with bad reputation rate is isolated. But in this approach reputation of node is not changed frequently, thus the nodes temporarily suffering from bad environmental conditions are not punished severely

2.2 Credit based scheme

Sprite Simple, cheat-proof, credit based system; here the node s send CAS (Central Authorized Server) a receipt for every packet they forward, CAS gives credits to nodes according to the receipt. This approach is useful as it is easy to implement but the major problem is scalability and message overhead.

Ad hoc-VCG(Vickery, Clarke and Groves) scheme ,this is a two phase approach in the Route Discovery phase destination node computes needed payments for intermediate nodes and notifies it to the source node or the central bank. In the Data Transmission phase actual payment is performed .This scheme is fully depends on the report of the destination node.

2.3 Game Theoretic scheme

In game theoretic scheme the IDS compares node's performance against other node based on a repeated game. This scheme is easy to implement but it needs fair comparison among nodes other wise it may falsely identify a node as adversary node.

3 MOTIVATIONS

The initial motivation for our work is to address limitations of current IDS systems by taking advantage of the mobile agent paradigm. Specifically, we address the following limitations of the earlier proposed IDS

False Positive Rate: The IDS reduces the False Positive rate that may arise in Reputation based scheme, which effectively increase the network performance.

Scalability: The process scalability of the credit based approach or any centralized approach is much lower. By using Mobile Agent the scalability may increase that enhance the network performance.

Interdependencies: In the Credit based scheme the IDS depends on the report of the destination node that make the network not convenient that require for MANET.

Centralized Authorization: Due to centralized authorization of previous IDS the IDS can not perform efficiently. In Mobile Agent based IDS the computation is done in distributed manner that increase the efficiency of the IDS.

4 PROPOSED WORK

Our objective is to find out the malicious node that performs the DOS by selfish node in network. The assumptions regarding the proposed work are listed below

4.1 Assumption

The following assumptions are taken in order to design the proposed algorithm.

1. A node interacts with its 1-hop neighbors directly and with other nodes via intermediate nodes using multi-hop packet forwarding.

2. Every node has a unique id in the network, which is assigned to a new node collaboratively by existing nodes.

3. The source node generates mobile agent after a specific period of time.

4. The mobile agent moves towards forward path created using RREQ and RREP.

5. The agent calculates the packet receive and forward by a node.

6. If the agent discovers a malicious node, instead of moving forward, it sends a report to the source node.

4.2 Architecture

Architecture of a Mobile agent based system:

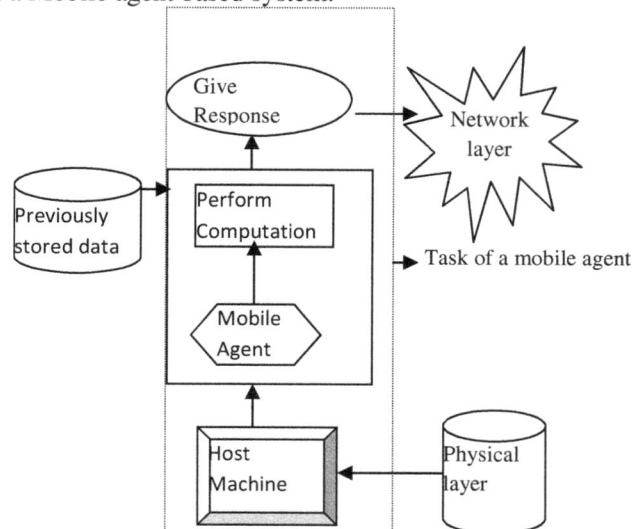

Fig 2: Architecture of proposed Mobile Agent IDS

From the above figure, it is observed that the mobile agent performs three tasks. At first the mobile agent (MA) has to collect the raw data from the host machine then it computes the packet delivery ratio (P_{dr}) after computation it compares the resultant P_{dr} with the predefined one and then gives responses to the source node accordingly.

The Mobile Agent maintains the following table to perform the computation and comparison with threshold value

Table 1: Data structure of the Mobile Agent

Source node ID	Destination Node ID	HOP count	THRESOLDP$_{dr}$

The table contains the source node id, destination node id that will be initiated by the source node. The HOP count field in the table denotes number of HOP between source node and destination node. THRESOLDP$_{dr}$ signifies the number of packet drop to be considered for any node in the forward path. The forward path is generated by the AODV routing protocol.

4.3 Methodology

The network is modeled based on the de-bruijn graph as follows:

Node Sequence: The Node sequence describes a set of nodes where the link among the nodes are created in such a way that when the node n with bit sequence (a0n a1n a2n.... akn) is connected with a node m having a bit sequence (a0m a1ma2m.... akm) where 1<=m,n<=r-1], then (ajm = ai n +1) where 1<=i,j<=k-1. Each node has in-degree and out-degree r. k is the diameter of the network represent as graph [12]. Here, the degree depends on the number of nodes in the forward path but for sake of simplicity the rank is always kept two. For a network where the number of nodes in the forward path including source and destination node is 7 the degree (d) should be computed as

C=7, r=2

We consider that d for which the following conditions are satisfied

1. (2d –C) is minimum
2. 2d >C

i) d=1 2d =2 2<7
ii) d=2 2d=4 4<7
iii) d=3 2d=8 8>7 and 2d-1=1
iv) d=4 2d=16 16>7 2d-1=9

For the first two computations 2nd condition is not satisfied for the 4th computation 1st condition is not satisfied so the degree is taken as 3.The digits are {0,1,2}

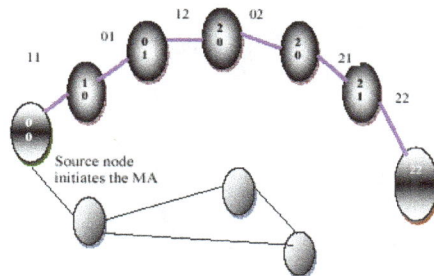

Fig 3: Nodes have unique ID compute by debrujin graph

Definition1: Packet Receive Counter: The number of packet that a node i receive from a neighbor node j is denoted as CPR (i,j)(Packet Receive Counter),1<=i,j<=N,where N is the total number of node in the network and i≠j and CPR (i,j)>=1.

Definition2: Packet Forward Counter: Total number of packet that a node i forward to its neighbors j is defined as CPF (i,j) (Packet Forward Counter) where 1<=i,j<=N-1 and i≠j.

Definition3: Packet Delivery Ratio (Pdr (i, j)): This is defined as the ratio of CPF (i, j) (Packet Forward Counter) of each node i for each neighbor j to the CPR (i, j) (Packet Receive Counter), 1<=i, j<=n and i≠j

$$CPR(i,j)= CPR(i,j)+1 \dots\dots\dots\dots\dots\dots\dots\dots\dots (1)$$
$$CPF (i,j) = CPF (i,j) + 1 \dots\dots\dots\dots\dots\dots\dots\dots(2)$$
$$Pdr(i,j)= CPF (i,j)/CPR(i,j) \dots\dots\dots\dots\dots\dots\dots\dots\dots\dots..(3)$$

If this Pdr(i,j)> THRESHOLDPdr(i,j) , It mark the ith neighbor as malicious node and inform source node

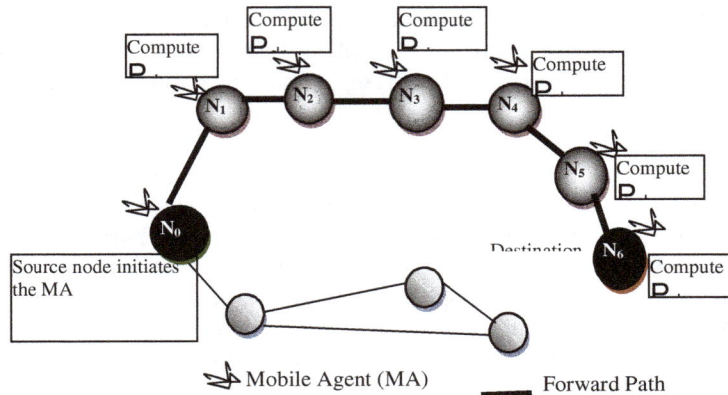

Fig 4: A network Without Malicious Attack The MA moves from source node N_0 to Destination Node N_6 by the forward path.

The figure 5 describes the situation when the network is under DoS attack by a selfish node. Here the node N2 sends acts as a malicious node RREQ messages to the node N2.The node N2 is a node in the forward path from source to destination node. N2 behaves as selfish node and refuse to forward packet to the neighbor node N3. When the MA comes to the node N2 it observes that the node behaves as malicious node by computing Pdr(N2,N3). This value is greater than THRESOLDPdr (N2, N3) and it send MMSG (Malicious Message) to the source node.

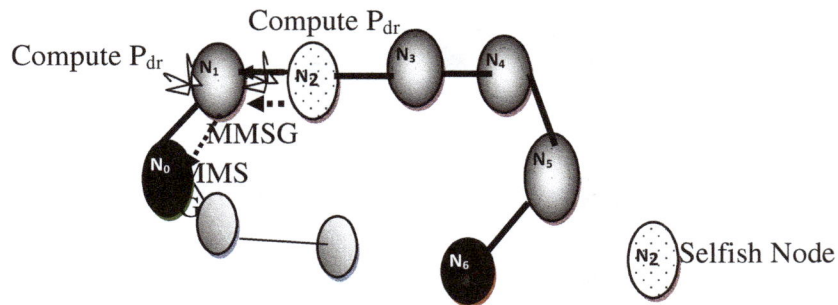

Fig 5: A network With the Malicious Attack

4.4 Algorithm:

In the figure 4 the source nodes N_0 generates the mobile agent and send it to the closest neighbor N_1. The MA at N_1 compute CPR (i, j) according to the equation 1. MA then calculates CPF (i, j) using equation 2, and then computes P_{dr} (i, j) using equation 3. If the Pdr (i, j) is greater than THRESHOLDP$_{dr}$ (i, j), then MA readily informs the source node via the intermediate nodes. From figure 4 it is observed that the MA reaches the destination node only when the network is free from DoS attack by a selfish node. The source uses the same path for others packets to be sent.

/* The following algorithm depicts the task of a mobile agent*/

Begin

        Step1: the source node N0 sends packet to the destination node N6

        Step2: Start Timer T

        Step3: Wait for the acknowledgement from destination node

        Step 4: increase T by unit time

        Step 5: if T>Tout then

            Goto step 6

          Else

            Goto step 3

        Step 6: The node S generates Mobile Agent(MA) and provides it's own ID and

            send it to the next hop node

        Step7: The mobile agent observe for ith node the number of packet receive from
           neighbor madoe j and compute CPR(i, j

        Step 8: MA compute CPF(i,j)for the ith node

        Step 9: MA compute Pdr(I,j) for the ith node at tth instance

        Step 10: If the ratio is less than threshold for ith node

           Then

               The agent moves to the next hop node

               decrase hop count by 1

          Else

               Agent reports the malicious activity to the source node

End

6 PERFORMANCE ANALYSES

6.1 Simulation Metric

Simulation metrics are the important determinants of network performance, which have been used to compare the performance of the proposed scheme in the network.

End to End Delay (D): The End to End Delay (D) is defined as the time of reception of the packet by the destination node (T_d) and the time of generation of the packet by the source node (T_s) for a sequence of packet P_{seq}.

For the packet sequence P^1_{seq} $D_1 = T^1_d - T^1_s$
For the packet sequence P^2_{seq} $D_2 = T^2_d - T^2_s$
For the packet sequence P^n_{seq} $D_n = T^n_d - T^n_s$

$$\text{Average End to End Delay} = \frac{\sum_{i=1}^{n} Di}{Npkt}$$

Where N_{pkt} is the total number of packet

Number of data packets Received: This parameter computes the total number of data packet received by any node in the forward path. If the number of received packets increases, the throughput would increase.

Cumulative Sum of Receiving Packet: This is defined as the sequence of partial sums of all packets received by the destination node.

$$CU_{Sum} = \sum_{i=1}^{n} N^i_{pkt}$$

Where n is the total number of packet sends at i^{th} instance to the destination node

6.2 Performance Evaluation

From figure 6, we observe that the performance of the network in presence of malicious node degrades than the network with mobile agent. Due to presence of mobile agent the network performance improves as the network is prevented from the malicious node. Initially the Average Throughput of Receiving Packet is same implies that he network is free from network at that time instant. As the packet size increases the throughput decreases means due to packet overhead the throughput decreases.

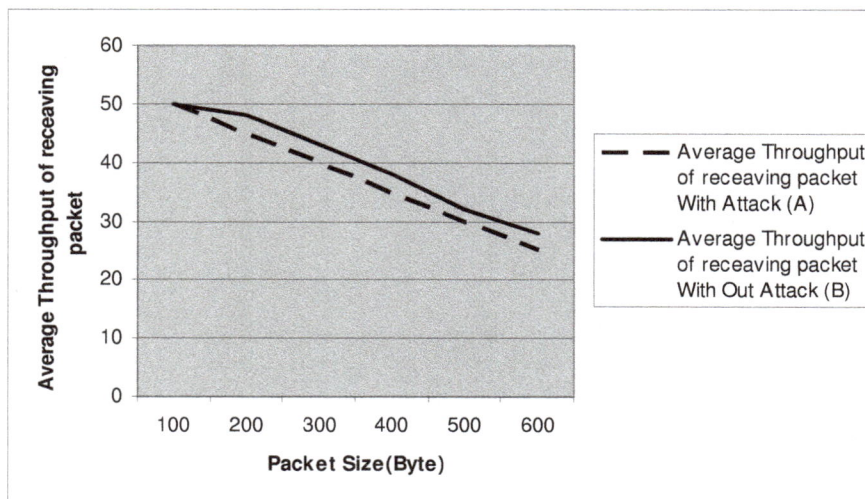

Fig 6 Average Throughput of Packet receive when the network is under attack and in presence of mobile agent

From figure 7, it is observed that cumulative sum of number of receive packet is more in presence of mobile agent. Series A describe the performance of the network with attack and series B describes the performance of network in presence of mobile agent.

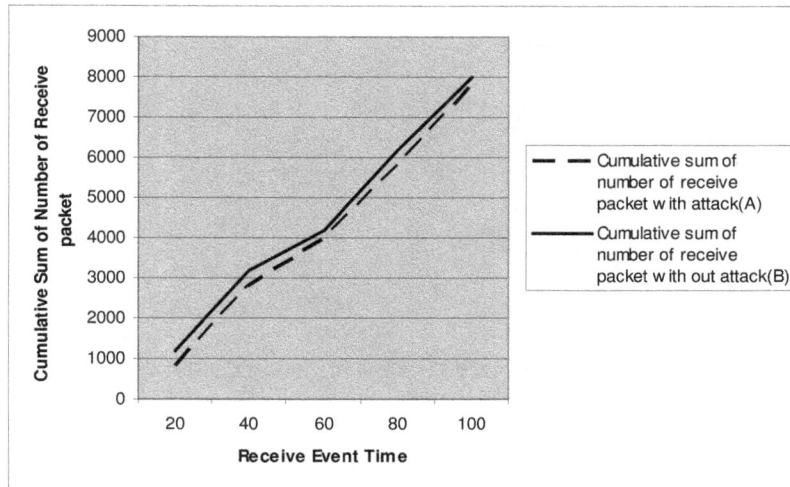

Fig 7 Cumulative Sum of number of Receive packet when the network is under attack and in presence of mobile agent

In figure 8, series "A" denotes average Throughput of Forwarding Packet from source to destination node in presence of malicious node attack in network. Series "B" indicates the Throughput of Forwarding Packet in presence of mobile agent in the network. At the simulation time instance 40sec the throughput is maximum indicating that at this moment the packet forward by the intermediate nodes are maximum in presence of malicious node. The source and the destination nodes are very close to each other in the network at this time instance.

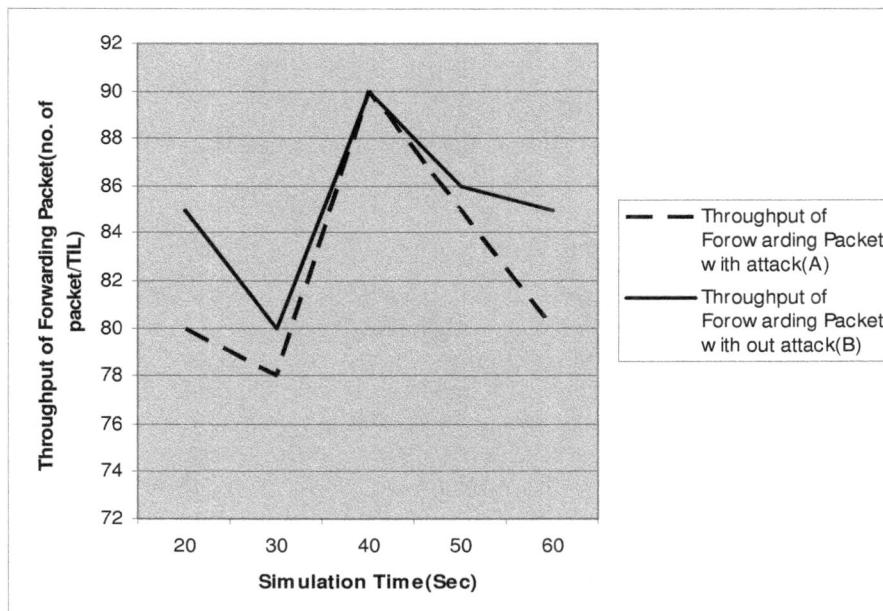

Fig 8 Throughput of forwarding packet when the network is under attack and in presence of mobile agent

From figure 9, we observe that the performance of the network in presence of selfish node degrades than the network without any attack. When the network is under attack in presence of mobile agent then the performance of the network remain same as that in case of the network without attack.

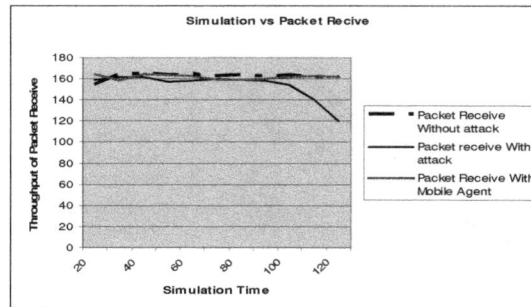

Fig 9 Throughput of Packet receive when the network is under attack and in presence of mobile agent

In figure 10, the series "a" indicates the average end to end delay in presence of DoS attack in the network. In series "b" the end to end delay increases as the packet size increases but the performance is better than that is shown in series "a". The pick of the graph denotes that at that point due to network congestion the delay is maximum.

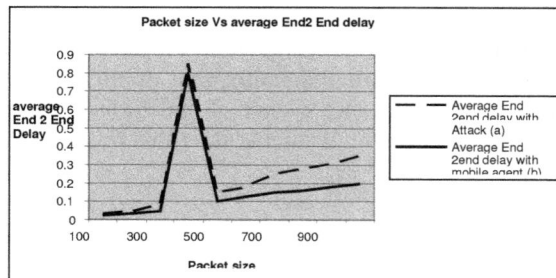

Fig 10 End 2 End Delay under attack and in presence of mobile agent

7 CONCLUSIONS AND FUTURE WORK

The mobile ad-hoc network suffers from several types of intrusions, out of which, the denial of service attack by a selfish node is the one of them. The mobile agents travel through the network, gathering vital information. This information is then processed by the mobile agent itself. The choice of threshold value is very important to help the detection of the attacker as early as possible. The computation complexity of the MA is kept minimum so that computation overhead can be reduced. In this paper we only focus on the DoS attack caused by selfish node by refusing the packet delivery to the neighbor node. The computation overhead of our algorithm is much less as the computation is done by the MA when the source node notices that the destination node does not response in correct time. The nodes are also free from performing the computation. This feature of our proposed scheme increases the efficiency of each node thereby increasing the overall performance of the network.

8 ACKNOWLEDGMENTS

This research is supported in part by the Computer Science and Engineering Department of University of Calcutta. We would like to thank Dr. Nabendu Chaki for fruitful discussions and endow with valuable suggestion.

REFERENCES

[1] A. S. Anand, M. Chawla, Detection of Packet Dropping Attack Using Improved Acknowledgement Based Scheme in MANET, IJCSI International Journal of Computer Science Issues, Vol. 7, Issue 4, No 1, July 2010

[2] T.V.P.Sundararajan1, Dr.A.Shanmugam2, Modeling the Behavior of Selfish Forwarding Nodes to Stimulate Cooperation in MANET, International Journal of Network Security & Its Applications (IJNSA), Volume 2, Number 2, April 2010

[3] 3. P.K.Suri and Kavita Taneja,Exploring Selfish Trends of MaliciousDevices in MANET In Journal Of Telecommunications, Volume 2, Issue 2, May 2010.

[4] Debdutta Barman Roy, Rituparna Chaki, Nabendu Chaki A New Cluster-Based Wormhole Intrusion Detection Algorithm for Mobile Ad-Hoc Networks, (IJNSA), Vol 1, No 1, April 2009

[5] Sukla Banerjee, Detection/Removal of Cooperative Black and Gray Hole Attack in Mobile Ad-Hoc Networks in Proceedings of the World Congress on Engineering and Computer Science 2008WCECS 2008.

[6] Xiaoxin Wu David K. Y. Yau,Mitigating Denial-of-Service Attacks in MANET by Distributed Packet Filtering: A Game-theoretic Approach, ASIACCS'07, 2007.

[7] Marko Jahnke, Jens Toelle, Alexander Finkenbrink,. Alexander Wenzel, et.al;Methodologies and Frameworks for Testing IDS in Adhoc Networks; Proceedings of the 3rd ACM workshop on QoS and security for wireless and mobile networks; Chania, Crete Island, Greece, Pages: 113 - 122, 2007

[8] Y.-C. Hu, A. Perrig, D. B. Johnson; Wormhole Attacks in Wireless Networks; IEEE Journal on Selected Areas of Communications, vol. 24, numb. 2, pp. 370-380, 2006

[9] Yang, H. and Luo, H. and Ye, F. and Lu, S. and Zhang, U.; Security in Mobile Ad Hoc Networks: Challenges and Solutions; Wireless Communications, IEEE, vol. 11, num. 1, pp. 38-47, 2004

[10] Y.-C. Hu, A. Perrig; A Survey of Secure Wireless Ad Hoc Routing; Security and Privacy Magazine, IEEE, vol. 2, issue 3, pp. 28-39, May 2004.

[11] Y.-C. Hu, A. Perrig, D. B. Johnson; Packet leashes: a defense against wormhole attacks in wireless networks; INFOCOM 2003, Twenty-Second Annual Joint Conference of the IEEE Computer and Communication Societies, Vol. 3, pp. 1976-1986, 2003

[12] Rituparna Chaki, Uma Bhattacharya dsign Of New Scalable Topology for Multihop Optical Network, IEEE TENCON 2000.

6

ACQUISITION PROBABILITY OF MULTI-USER UWB SYSTEMS IN THE PRESENCE OF A NOVEL SYNCHRONIZATION APPROACH

Moez Hizem[1] and Ridha Bouallegue[1]

[1]Innov'Com Laboratory, Sup'Com, University of Carthage, Tunis, Tunisia

moezhizem@yahoo.fr
ridha.bouallegue@supcom.rnu.tn

ABSTRACT

In this paper, to synchronize Ultra Wideband (UWB) systems in ad-hoc multi-user environments, we propose a new timing acquisition approach for achieving a good performance despite the difficulties to get there. Synchronization constraints are caused by the ultra-short emitted waveforms nature of UWB signals. Used in [1, 2] for single-user environments, our timing acquisition approach is based on two successive stages or floors. Extended for multi-user environments, the used algorithm is a combination between coarse synchronization based on timing with dirty templates (TDT) acquisition scheme and a new fine synchronization scheme developed in [3-6] which conduct to an improved estimate of timing offset. In this work, we develop and test this method in both data-aided (DA) and non-data-aided (NDA) modes. Simulation results and comparisons are also given to confirm performance improvement of our approach (in terms of mean square error and acquisition probability) compared to the original TDT algorithm in multi-user environments, especially in the NDA mode.

KEYWORDS

UWB; Time Hopping (TH); pulse amplitude modulation (PAM); synchronization; performance; mean square error (MSE) & acquisition probability

1. INTRODUCTION

UWB systems have received a great attention in the last years. Furthermore, the interest for commercial UWB technology is growing fast especially in the areas of high-data rate short-range wireless multimedia applications, as well as for low-data rate sensor networks [7]. One of these benefits comes from the large number of users allowed access with Time Hopping (TH) codes, and potential to overlay existing narrowband systems such as IEEE 802.11 and Bluetooth [8]. Hence, the idea of this work is to extend the timing acquisition approach previously developed in [1, 2] for multi-user environments this time.

Nevertheless, to exploit these benefits, one of the most important challenges (at least at the physical layer) is to obtain an accurate timing synchronization and more specifically timing offset estimation. In general, synchronization is usually achieved in two stages [9]. The first stage realizes coarse synchronization to within a realistic amount of precision in a short time, and is well-known as the acquisition stage. For this, we use the Timing with Dirty Templates (TDT) acquisition scheme introduced in [10] and developed in [11-14]. The TDT approach is an attractive technique for UWB systems, which is specified d by its low complexity and rapid acquisition in the DA mode. It's based on correlating adjacent symbol-long segments of the received waveform.

To improve the acquisition probability and the performance of multi-user UWB systems, we propose in the second stage a novel fine synchronization approach known as tracking stage which is responsible for preserving synchronization through clock drifts tacking place in the transmitter and the receiver. Tracking is usually accomplished with a delay locked loop (DLL) [9]. Timing acquisition is an enormous difficulty caused by UWB systems. Compared with the original TDT synchronizer, simulation results show that our new based-TDT-synchronizer can achieve a higher acquisition probability than the original TDT in both NDA and DA modes for multi-user environments.

The rest of this paper is organized as follows. The ensuing Section 2 describes the UWB TH-PAM system in multi-user environments. The Section 3 outlines our novel acquisition algorithm based on two stages. In Section 4, the simulations are carried out to corroborate our analysis in comparison with the original TDT approach. And finally, the conclusions are given in Section 5.

2. SYSTEM MODEL FOR MULTI-USER LINKS

The UWB time hopping impulse radio signal considered in this paper is a stream of narrow pulses, which are shifted in amplitude modulated (PAM). The same modulated pulse is repeated N_f times (frames number) over a T_s period (symbol time). During each duration frame T_f, a data-modulated ultra-short pulse p(t), with duration $T_p \ll T_f$, is transmitted [4]. The transmitted waveform from the uth user is

$$v_u(t) = \sqrt{\varepsilon_u} \sum_{k=0}^{+\infty} s_u(k) p_{u,T}(t - kT_s) \tag{1}$$

where ε_u represents the energy per pulse, $s_u(k)$ are differentially encoded symbols and drawn equiprobably from finite alphabet. In our case, $s_u(k)$ symbolize the binary PAM information symbols and $p_{u,T}(t)$ indicates the transmitted symbol as

$$p_{u,T}(t) := \sum_{i=0}^{N_f-1} p(t - iT_f - c_u(i)T_c) \tag{2}$$

where T_c is the chip duration and $c_u(i)$ is the user-specific pseudo-random TH code during the ith frame.

The transmitted signal propagates through the multipath channel corresponding to each user. The UWB channel is modelled as tapped-delay line with L_u taps, where $\{\alpha_{u,l}\}_{l=0}^{L_u-1}$ and $\{\tau_{u,l}\}_{l=0}^{L_u-1}$ is amplitude and delay of the L multipath elements, respectively. The channel is assumed quasi-static and among $\{\tau_{u,l}\}_{l=0}^{L_u-1}$, τ_0 represents the propagation delay of the channel. Thus, the received waveform from all users is

$$r(t) = \sum_{u=0}^{N_u-1} \sqrt{\varepsilon_u} \sum_{l=0}^{L_u-1} \alpha_{u,l} v_u(t - \tau_{u,l} - \tau_u) + \eta(t) \tag{3}$$

where N_u is the users number, τ_u is the propagation delay of the uth user's direct path and $\eta(t)$ is the zero-mean additive Gaussian noise (AGN). The global received symbol-long waveform is therefore given by

$$p_{u,R}(t) := \sum_{l=0}^{L_u-1} \alpha_{u,l} p_{u,T}(t - \tau_{u,l}) \tag{4}$$

Assuming that the nonzero support of waveform $p_{u,R}(t)$ is upper bounded by the symbol time T_s, the received waveform in (3) can be rewritten as

$$r(t) = \sum_{u=0}^{N_u-1} \sqrt{\varepsilon_u} \sum_{k=0}^{+\infty} s_u(k) p_{u,R}(t - kT_s - \tau_u) + \eta(t) \quad (5)$$

In the next section, we will develop a low-complexity timing acquisition approach using TDT synchronizer in order to find the desired timing offset for multi-user environments. The structure of our synchronization scheme is illustrated in Fig.1.

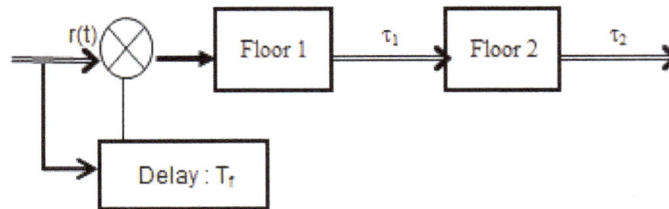

Figure 1. Structure of our synchronization scheme

3. PROPOSED TIMING ACQUISITION APPROACH

There are many difficulties reaching from the signal and channel characteristics relied to UWB systems that show the importance of the timing acquisition problem and the necessity to resolve it effectively, in particular for multi-user environments. This signifies that there could be numerous steps in the research domain which may be judged satisfactory and may be developed to accelerate the timing acquisition processing.

As mentioned previously, the new acquisition timing scheme proposed in this paper for multi-user environments consists of two complementary stages or floors. The first one is based on a blind (or coarse) synchronization algorithm which id the well-known TDT. The system model structure's with first stage synchronization is shown in Fig.2. This approach is consisted on correlating adjacent symbol-long segments of the received waveform. The basic idea behind TDT is trying to find the maximum of square correlation between pairs of successive symbol-long segments. These symbol-long segments are also called dirty templates and they are subject to the unknown offset τ_0. Then, we will analyze $\hat{\tau}_0'$ representing estimate offset of τ_0 by deriving upper bounds on their mean square error (MSE) in both NDA and DA modes.

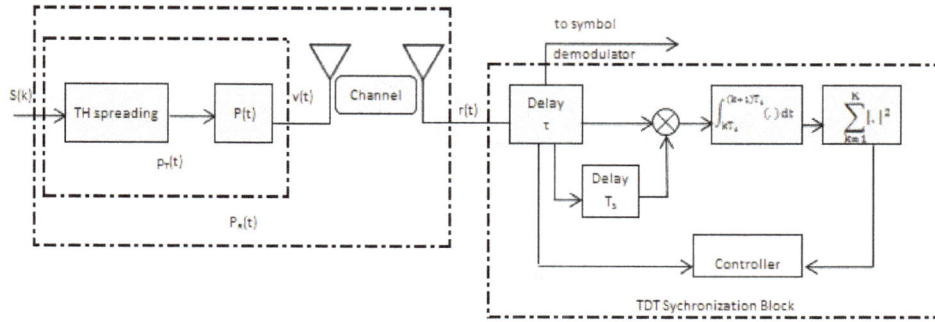

Figure 2. System model structure's with first stage synchronization

For multi-user UWB TH-PAM systems, a correlation between the two adjacent symbol-long segments $r(t - kT_s)$ and $r(t - (k - 1)T_s)$ is achieved. Let $x(k; \tau)$ the value of this correlation $\forall k \in [1, +\infty)$ and $\tau \in [0, T_s)$,

$$x(k; \tau) = \sum_{u=0}^{N_u-1} \int_0^{T_s} r(t - kT_s) r(t - (k - 1)T_s) \, dt \qquad (6)$$

After development and calculation, we find the above equation as,

$$x(k; \tau) = \chi_0(k; \tau) + \sum_{u \neq 0} s_u(k - 1)\left[s_u(k)\varepsilon_{u,B}(\check{\tau}_u) + s_u(k - 2)\varepsilon_{u,A}(\check{\tau}_u)\right] + \xi(k; \tau) \qquad (7)$$

where $s_u(k)$'s are zero-mean information symbols emitted by the $(u \neq 0)$th user, $\varepsilon_{u,A}(\check{\tau}_u) := \varepsilon_u \int_{T_s-\check{\tau}_u}^{T_s} p_{u,R}^2(t) \, dt$, $\varepsilon_{u,B}(\check{\tau}_u) := \varepsilon_u \int_0^{T_s-\check{\tau}_u} p_{u,R}^2(t) \, dt$, $\check{\tau}_u := [\tau_u - \tau]_{T_s}$ and $\xi(k; \tau)$ corresponds to the superposition of three noise terms [10] and can be approximated as an additive white Gaussian noise (AWGN) with zero mean and σ_ξ power.

Practically, the mean square of $x^2(k;\tau)$ is approximated from the average of different values $x^2(k;\tau)$ for k ranging from 0 to M–1 acquired during an observation interval duration's MT_s. In the multi-user case, for the two synchronization modes NDA and DA, the TDT algorithm is given as follows,

$$\hat{\tau}_u = argmax_{\tau \in [0,T_s]} E\left\{x^2(k;\tau)\right\}$$

$$x(M; \tau_u) = \frac{1}{M} \sum_{m=0}^{M-1} \left(x(k; \tau)\right)^2 \qquad (8)$$

As a second stage, we propose a new fine synchronization approach with low complexity in order to have a more accurate estimate of the exact time synchronization. In this stage, we analyze and develop a fine synchronization algorithm that will give a better estimate of the time delay. The second stage synchronization description's is given in Fig 3.

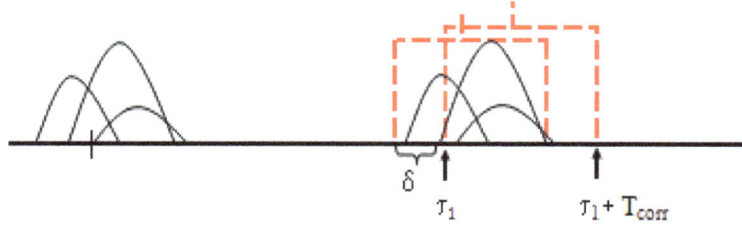

Figure 3. Description of second synchronization stage

This floor realizes a fine estimation of the frame beginning, after a blind research in the first. The idea which is based this floor is extremely simple. The proposition is to scan the time period $[\tau_1 - T_{corr}, \ \tau_1 + T_{corr}]$ through a step noted δ by making integration between the received signal and its replica shifted by T_f on a window with a width T_{corr}. τ_1 is the estimate delay found after the first synchronization stage. We denote the integration window output for the nth step $n\delta$ as follows

$$Z_n = \sum_{k=0}^{K-1} \left| \int_{\tau_1 + n\delta}^{\tau_1 + n\delta + T_s} r(t - kT_s)r(t - (k+1)T_s)dt \right| \qquad (9)$$

where $n = -N + 1..0..N - 1$, $N = \lfloor T_{corr}/\delta \rfloor$ and K is the frames number considered for enhancing the decision taken at the first stage. The value of n which maximizes Z_n gives the exact moment of pulse beginning noted $\tau_2 = \tau_1 + n_{opt}\delta$. Consequently, the fine synchronization is achieved. Finally, note that this proposed timing acquisition approach will be developed in both NDA and DA modes. In the next Section, we will deduce for multi-user environments in what mode this approach gives us better results compared to those given by the original approach TDT in terms of mean square error (MSE) and especially acquisition probability.

4. SIMULATION RESULTS AND COMPARISONS

In this section, we will evaluate the performance of our timing acquisition approach in multi-user environments with simulations. The UWB pulse is the second derivative of the Gaussian function with unit energy and duration $T_p \approx 0.8ns$. Simulations are achieved in the IEEE 802.15.3a channel model CM1 [15]. The sampling frequency chosen in the simulations is $f_c = 50$ GHz. Each symbol contains $N_f = 32$ frames each with duration $T_f = 35$ ns. We used a random TH code uniformly distributed over $[0, N_c - 1]$, with $N_c = 35$ and $T_c = 1.0$ ns. The width integration window value's T_{corr} is 4 ns. The performance of our approach is tested for various values of M with the presence of two interfering users. The two interfering users are asynchronous relative to the desired user, and are sending information symbols with 5 and 10 dB less SNR than the desired one.

In Figs. 4-5, we compare the multi-user performances in terms of mean square error (MSE) of both original TDT and fine synchronization approaches for different values of M [4]. From the simulation results, we note that increasing the duration of the observation interval M leads to improved performance for both NDA and DA modes. In comparison with the original TDT approach, we show also that the new timing acquisition approach greatly outperforms the NDA mode and offers a slight improvement in DA mode.

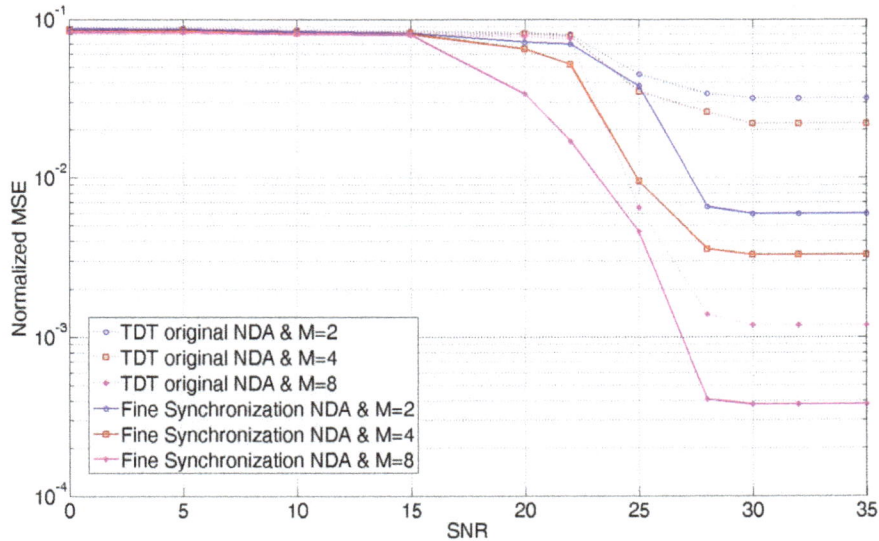

Figure 4. MSE Performances comparison in NDA mode with multi-user environments

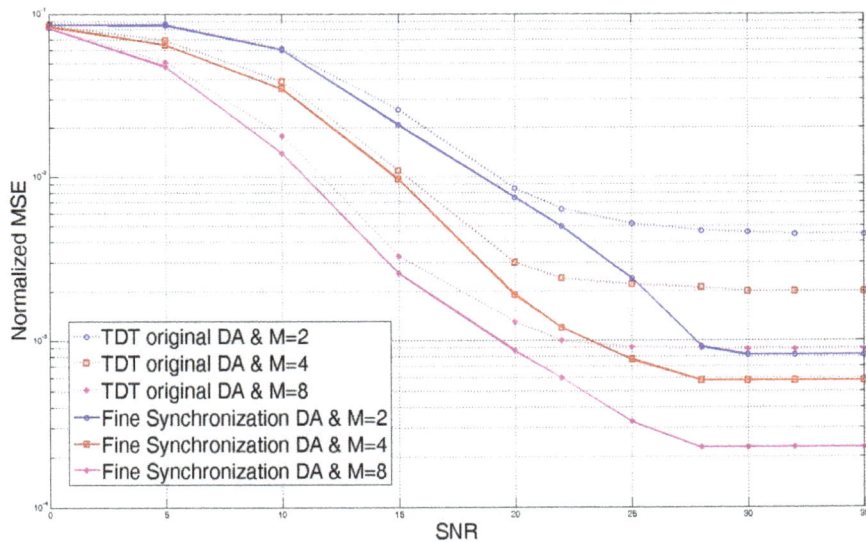

Figure 5. MSE Performances comparison in DA mode with multi-user environments

In Figs. 6-7, we compare the timing acquisition probability of the proposed synchronizer with the original TDT algorithm. From the simulation results, we deduce that our novel synchronizer can ameliorate the original NDA TDT algorithm and realize a slight improvement performance's to the DA TDT algorithm. This performance amelioration is enabled thanks to the contribution of fine synchronization approach introduced in second stage which can further improve the timing offset found in first stage (coarse synchronization approach: TDT). Unfortunately, this performance amelioration is permitted at the price of higher computation complexity and consequently time lost.

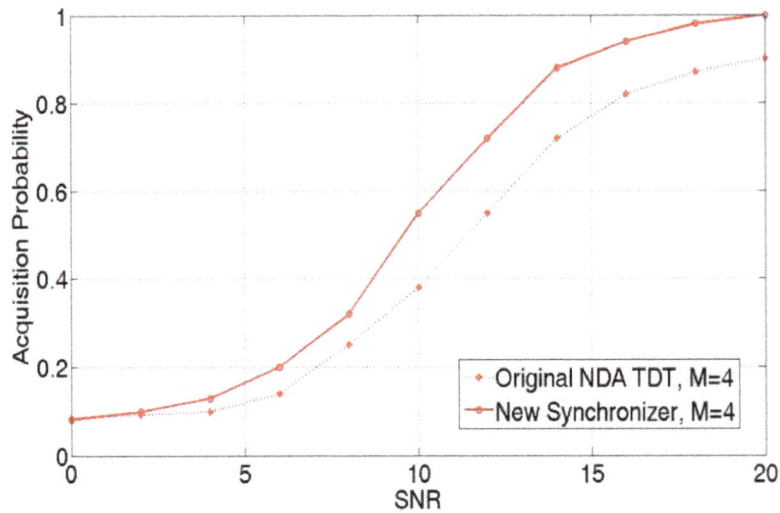

Figure 6. Multi-user acquisition probability comparison in NDA mode: proposed synchronizer vs. original TDT

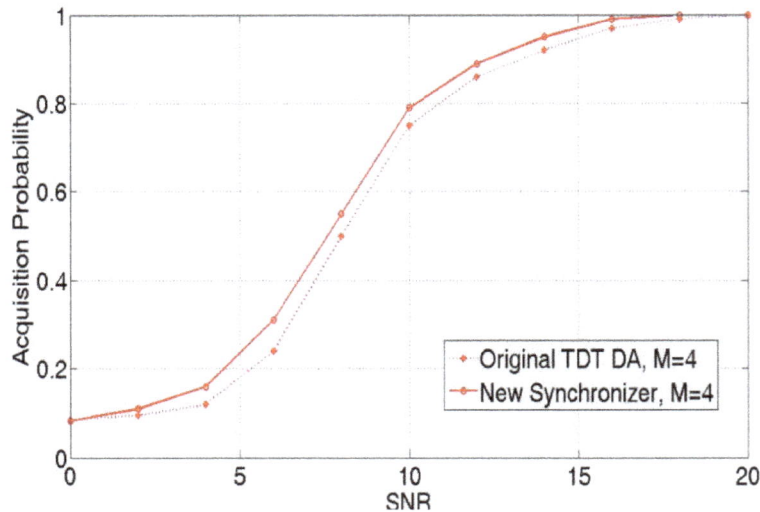

Figure 7. Multi-user acquisition probability comparison in DA mode: proposed synchronizer vs. original TDT

5. CONCLUSIONS

In this paper, we establish a new timing acquisition approach based on the timing with dirty templates (TDT) introduced in first stage of our synchronization algorithm for UWB radio system. With the fine synchronization approach introduced in second stage, we realize a fine estimation of the frame beginning. The simulation results show that even without training symbols, our novel synchronizer can enable a better performance (in terms of mean square error and acquisition probability) than the original TDT especially in NDA mode and offers a slight improvement in DA mode.

REFERENCES

[1]   M. Hizem, and R. Bouallegue, (2012) "Novel timing acquisition approach for UWB systems," in Proceedings of International Conference on Information Processing and Wireless Systems (IPWiS), Sousse, Tunisia, March 16-18.

[2]   M. Hizem, and R. Bouallegue, (2012) "UWB systems performance with a new timing acquisition approch," in Proceedings of International Symposium of Wireless and Communication Systems (ISWCS), Paris, France, August 28-31.

[3]   M. Hizem, and R. Bouallegue, (2010) "Novel Fine Synchronization Using TDT for Ultra Wideband Impulse Radios," in Proceeding of International Information and Telecommunications Technologies Symposium (I2TS), Rio de Janeiro City, Rio de Janeiro State, Brazil, Dec. 11-13.

[4]   M. Hizem, and R. Bouallegue, (2011) "Fine Synchronization with UWB TH-PAM Signals in ad-hoc Multi-user Environments," in Proceeding of Progress in Electromagnetics Research Symposium (PIERS), Marrakech, Morocco, March 20-23.

[5]   M. Hizem, and R. Bouallegue, (2011) "Fine Synchronization through UWB TH-PPM Impulse Radios," in Proceeding of International Journal of Wireless & Mobile Networks (IJWMN), Vol.3, No.1, February.

[6]   M. Hizem, and R. Bouallegue, (2011) Fine Synchronization in UWB Ad-Hoc Environments, book chapter published in Novel Applications of the UWB Technologies, ISBN 978-953-307-324-8, Publisher: InTech, Edited by: Boris Lembrikov, Publication date: August 2011, pp. 123-140.

[7]   L. Yang and G. B. Giannakis, (2005) "Ultra-wideband communications: an idea whose time has come," IEEE Communications Magazine, vol. 43, no. 2, pp. 80–87.

[8]   M. Z. Win and R. A. Scholtz, (2000) "Ultra wide bandwidth time-hopping spread-spectrum Impulse Radio for wireless multiple access communications," IEEE Trans. on Communications, vol. 48, no. 4, pp. 679–691.

[9]   R. L. Peterson, R. E. Ziemer, and D. E. Borth, (1995) An Introduction to Spread Spectrum Communications. Upper Saddle River, NJ: Prentice-Hall.

[10]   L. Yang and G. B. Giannakis, (2005) "Timing UWB signals using dirty templates," IEEE Transactions on Communications, Vol.53, no.11, pp. 1952-1963.

[11]   L. Yang, (2006) "Timing PPM-UWB signals in ad hoc multi-access," IEEE Journal on Selected Areas in Communications, vol. 24, no. 4, pp. 794-800.

[12]   W. Zang, W. Zhang, and L. Yang, (2009) "On the Optimality of Timing with Dirty Templates," IEEE Global Telecommunications (GLOBECOM), Honolulu, HI.

[13]   H. Xu, and L. Yang, (2008) "Timing with Dirty Templates for Low-Resolution Digital UWB Receivers," IEEE Transactions on Wireless Communications, vo.7, no.1, pp. 54-59.

[14]   M. Ouertani, H. Xu, L. Yang, H. Besbes, and A. Bouallegue, (2007) "A New Modulation Scheme For Rapid Blind Timing Acquisition Using Dirty Template Approach for UWB Systems," IEEE International Conference on Acoustics, Speech and Signal Processing (ICCASP), Honolulu, HI, pp. 265-268.

[15]   J. R. Foerster, (2002) "Channel Modeling Sub-committee Report Final," *IEEE P802.15-02/368r5-SG3a, IEEE P802.15 Working Group for WPAN.*

7

PERFORMANCE ANALYSIS OF DSDV AND DSR UNDER VARIABLE NODE SPEED IN HYBRID SCENARIO

Koushik Majumder[1], Sudhabindu Ray[2] and Subir Kumar Sarkar [2]

[1] Department of Computer Science & Engineering, West Bengal University of Technology, Kolkata, INDIA
koushik@ieee.org
[2] Department of Electronics and Telecommunication Engineering, Jadavpur University, Kolkata, INDIA

ABSTRACT

The mobile ad hoc networks have gained immense popularity in the current decade due their less costly and rapid deployability, inherent support for mobility and the potential to provide ad hoc connectivity to devices. Routing in mobile ad hoc network is considered as a challenging task due to the drastic and unpredictable changes in the network topology resulting from the random and frequent movement of the nodes and due to the absence of any centralized control. Routing becomes even more complex in hybrid networking scenario where the MANET is combined with the fixed network for covering wider network area with less fixed infrastructure. Although, several routing protocols have been developed and tested under various network environments, but, the simulations of such routing protocols have not taken into account the hybrid networking environments. In this work we have carried out a systematic simulation based performance study and analysis of the two prominent routing protocols: Destination Sequenced Distance Vector Routing (DSDV) and Dynamic Source Routing (DSR) protocols in the hy-brid networking environment under varying node speed. We have analyzed the performance differentials on the basis of three metrics – packet delivery fraction, average end-to-end delay and normalized routing load using NS2 based simulation.

KEYWORDS

Mobile ad hoc network, hybrid network scenario, varying node speed, performance analysis, packet delivery fraction, average end-to-end delay, normalized routing load.

1. INTRODUCTION

The mobile ad hoc networks(MANET) [1-14] have received increased attention of the research community in the current decade due their self organizing, self controlled and distributed nature of operations which separate them from the fixed networks.The main advantage of these networks is that their non-reliance on any established infrastructure or centralized server. These networks are autonomous where a number of mobile nodes equipped with wireless interfaces communicate with each other either directly or through other nodes. These networks are especially useful in emergency scenarios where there is no fixed infrastructure or the previous infrastructure is totally destroyed and it is not possible to set up a new infrastructure quickly. The communication in MNAET is multi-hop and each node has to play the role of both the host as well as the router. But due to the limited transmission range of the MANET nodes, the total area of coverage is often limited. Also due to the lack of connectivity to the fixed network, the users in the MANET work as an isolated group. However, many applications require connection to the external network such as Internet or LAN to provide the users with external resources.

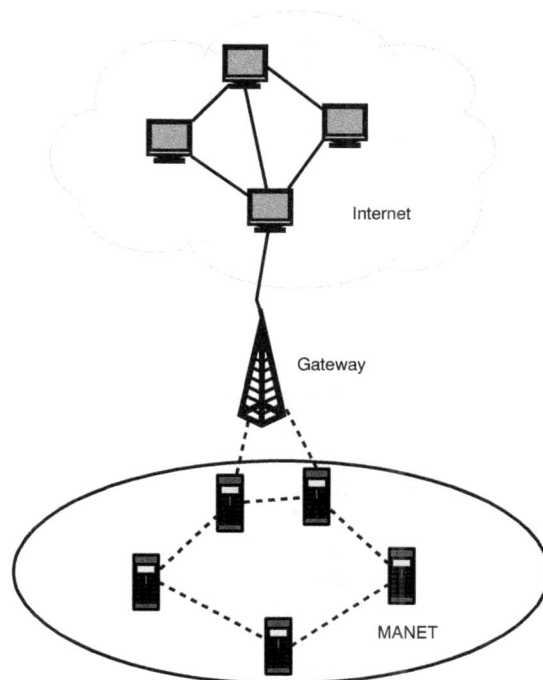

Figure 1. Hybrid Network

On the other hand the growth of Internet has been tremendous in the current decade and with its reducing cost of use, it has occupied a huge part of the lives of the common people. For example people present in any part of the world and connected to the Internet can communicate between them in almost no time using email, online audio and video chat. This plays an important role in the field of academics and research. No longer are the students restricted by their physical presence. For example students in distant areas can participate in the online classroom facilities provided by the top universities, they can download the study materials, raise questions and discuss their problems online. The researchers and scientists in different parts of the world can collaborate and work in groups and exchange their ideas instantaneously. Their distributed geographic presence no longer constrains the scope and rapid growth of research. On the other hand with the huge influx of mobile phones, laptops and personal digital assistants along with their reduced cost, mobility has become an indispensable part of our daily lives. These devices are highly portable and can be carried anytime anywhere. With the increasing use of these devices there is a growing demand for the connectivity to the Internet while we are on the move.

In order to access the global services and applications of the Internet and for widening the coverage area of the MANET, sometimes a hybrid network can be formed by combining the ad hoc network with the wired network. By using this combination we can cover a larger area with less fixed infrastructure, less number of fixed antennas and base station and can reduce the overall power consumption. Due to the hybrid nature of these networks, routing is considered a challenging task. Several routing protocols have been proposed and tested under various traffic conditions. However, the simulations of such routing protocols have not taken into account the hybrid network scenario. In this work we have carried out a systematic performance study of the two prominent routing protocols: Destination Sequenced Distance Vector Routing (DSDV) and Dynamic Source Routing (DSR) protocols in the hybrid networking environment under different node speed.

The rest of the paper is organized as follows. Section 2 describes the related work. A brief introduction of Dynamic Source Routing (DSR) and Destination Sequenced Distance Vector Routing (DSDV) protocols is given in Section 3. Section 4 and section 5 details the simulation model and the key performance metrics respectively. The simulation results are presented and analyzed in section 6. Finally the conclusion has been summarized in the section 7. The last section gives the references.

2. RELATED WORK

Several simulation based experiments have been made to compare the performance of the routing protocols for mobile ad hoc network.

Das et al. [15] made performance comparison of routing protocols for MANET based on the number of conversations per mobile node for a given traffic and mobility model. Small networks consisting of 30 nodes and medium networks consisting of 60 nodes were used. Simulation was done using the Maryland Routing Simulator (MARS).

Performance comparison results of two on demand routing protocols – AODV and DSR is presented in the work of Das, Perkins and Royer [16]. They used NS2 based simulation. CBR sources were used with packet size of 512 bytes. Two different simulation set ups were used. One with 50 nodes and 1500m x 300m simulation area and the other with 100 nodes and 2200m x 600m simulation area. The performance metrics studied were: packet delivery fraction, average end-to-end delay and normalized routing load.

Johansson, Larssson, Hedman and Mielczarek [17] in their work incorporated new mobility models. A new mobility metric was introduced to characterize these models. Using this metric, mobility was measured in terms of relative speeds of the nodes instead of absolute speeds and pause times. The network consisted of 50 nodes. There were 15 sources and the data packets transmitted were of 64 bytes. Performance analysis was made in terms of throughput, delay and routing load.

Park and Corson [18] made a performance comparison between TORA and an "idealized' link state routing protocol. Many simplifications were made in the simulation environment. For example, in the simulation scenario packets were transmitted at the rate of only 4, 1.5, or 0.6 packets per minute per node for avoiding congestion. Total duration of the simulation run was 2 hours. The network was connected in a "honeycomb" pattern. The node density was kept constant artificially. The notion of true node mobility was missing. Every node was connected to a fixed set of neighboring nodes through separate links. Each link switched between active and inactive states irrespective of other links. Immediate feedback was available when a link went up or down which is not the case in reality.

These works, however, do not take into consideration the influence of hybrid network scenario over the performance of the routing protocols. In this work we have studied the effect of varying node speed on the performance of two prominent routing protocols for mobile ad hoc network – Destination Sequenced Distance Vector Routing (DSDV) and Dynamic Source Routing (DSR) protocol in the hybrid networking environment.

3. DESCRIPTION OF ROUTING PROTOCOLS

3.1. Dynamic Source Routing (DSR)

The Dynamic Source Routing Protocol (DSR) is a reactive routing protocol. The main feature of DSR is the use of source routing technique. In this technique the source node knows the complete hop-by-hop route towards the destination node. The source node lists this entire sequence in the packet's header. If a node wants to send a packet to a destination, the route to which is unknown, in that case a dynamic route discovery process is initiated to discover the route. DSR consists of the Route Discovery and Route Maintenance phase, through which it discovers and maintains source routes to arbitrary destinations in the network.

3.1.1. Route Discovery

If a node A wants to send a packet to a destination node B, it searches its Route Cache. If the Route Cache contains a valid route, node A inserts this route into the header of the packet and sends the data packet to the destination B. In case when no route is found in the Route Cache, a Route Discovery is initiated.

Node A initiates the Route Discovery by broadcasting a ROUTE REQUEST message. All nodes within the transmission range receive this message. The nodes which are not in the route, add their address to the route record in the packet and forward the packet when received for the first time. They check the request id and source node id to avoid multiple retransmissions. The destination node B sends a ROUTE REPLY when it receives a ROUTE REQUEST. If the link is bidirectional, the ROUTE REPLY propagates through the reverse route of the ROUTE REQUEST. If the link is unidirectional, in that case B checks its own Route Cache for a route to A and uses it to send the ROUTE REPLY to the source A. If no route is found, B will start its own Route Discovery. In order to avoid infinite numbers of Route Discoveries it piggybacks the original ROUTE REQUEST message to its own. The route information carried back by the ROUTE REPLY message is cached at the source for future use. In addition to the destination node, other intermediate nodes can also send replies to a ROUTE REQUEST using cached routes to the destination.

3.1.2. Route Maintenance

The node which sends a packet using a source route is responsible for acknowledging the receipt of the packet by the next node. A packet is retransmitted until a receipt is received or the maximum number of retransmissions is exceeded. If no confirmation is received, the node transmits a ROUTE ERROR message to the original sender indicating a broken link. The ROUTE ERROR packet causes the intermediate nodes to remove the routes containing the broken link from their route caches. Ultimately the sender will remove this link from its cache and look for another source route to the destination in its cache. If the route cache contains another source route, the node sends the packet using this route. Otherwise, it needs to initialize a new route discovery process. DSR makes very effective use of source routing and route caching. In order to improve performance any forwarding node caches the source route contained in a packet forwarded by it for possible future use.

3.2. Destination Sequenced Distance Vector Routing (DSDV)

The Destination Sequenced Distance Vector Routing (DSDV) is a proactive or table driven routing protocol designed for MANET. It was developed by C. Perkins and P. Bhagwat in 19994. This scheme is based on the classical Bellman-Ford distance vector algorithm with certain modifications to make it suitable for the ad hoc environment and to solve the problem of routing loop and count-to-infinity. In DSDV every node maintains a routing table which contains the list of all possible destinations within the network and the number of hops to reach each possible destination. Each distance entry is marked by a sequence number usually originated by the destination node. This sequence numbering scheme is used to counter the count-to infinity problem and to distinguish the stale routes from the fresh ones thus avoiding the formation of loops.

In order to maintain up-to-date routing information about the frequently changing topology of the network the nodes need periodic exchanges of routing tables with their neighbours. But this will create a huge overhead of control packets in an already bandwidth constrained network. To reduce this huge overhead of control traffic the routing updates are generally classified into two types – full dump and incremental update. In case of full dumps, nodes need to exchange complete routing tables with their neighbours. Full dumps are needed to maintain consistent routing information when the network topology changes completely and very fast due to frequent movement of nodes. But this may result in a large number of routing packet exchanges between the nodes. On the other hand incremental updates contain only those entries that have been updated since the last full dumps. Incremental updates are much smaller in size than the full dumps and they should fit in a single Network Protocol Data Unit (NPDU). When the network is relatively stable, incremental updates are used to rapidly propagate the routing information regarding the small changes in network topology. This saves a lot of network traffic. In addition to the periodic updates DSDV uses triggered updates, when significant new information is available about the topological change. Thus the update is both time-driven as well as event-driven.

Table updates are initiated by the destination nodes and they generate the sequence numbers. Every node periodically transmits their routing updates to their immediate neighbours with monotonically increasing sequence numbers. After receiving a new route update, every node compares it with its existing entry. Routes with smaller sequence numbers are simply discarded and the one with the recent sequence number is selected. In case when the new route is having the same sequence number as the existing route, the one with the smaller hop count is selected. If the new route is chosen, its hop-count is incremented by one, as the packets will require one more hop to reach the destination. This change in the routing information is then immediately communicated to the neighbours.

When a node S finds that its route to destination D is broken, it advertises its link to destination D with an infinite hop-count and a sequence number that is one greater than the sequence number of the broken route. This is the only case when the sequence number is not assigned by the destination node. Sequence numbers defined by the originating nodes are even numbers, whereas the sequence numbers indicating the broken links are odd numbers. After having this infinite hop-count entry, when a node, later receives a finite hop-count entry with newer sequence number, it immediately broadcasts its new routing update. The broken links are thus quickly replaced by the real routes.

4. SIMULATION MODEL

We have done our simulation based on ns-2.34 [19-22]. NS is a discrete event simulator. It was developed by the University of California at Berkeley and the VINT project [19]. Our main goal was to measure the performance of the protocols under a range of varying network conditions. We have used the Distributed Coordination Function (DCF) of IEEE 802.11[23] for wireless LANs as the MAC layer protocol. Data packets were transmitted using an unslotted carrier sense multiple access (CSMA) technique with collision avoidance (CSMA/CA) [23].

The protocols have a send buffer of 64 packets. In order to prevent indefinite waiting for these data packets, the packets are dropped from the buffers when the waiting time exceeds 40 seconds. The interface queue has the capacity to hold 80 packets and it is maintained as a priority queue. The interface queue holds both the data and control traffic sent by the routing layer until they are transmitted by the MAC layer. The control packets get higher priority than the data packets.

4.1. Mobility Model

Inclusion of a mobility model is necessary in order to evaluate the performance of a protocol for ad hoc network in a simulated environment. Here in our work we have used the random waypoint model. This model is a simple and common mobility model and is widely used for the performance evaluation of MANET protocols in simulated environment. This particular mobility model has pause time between changes in direction and/or speed. The mobile nodes are initially distributed over the entire simulation area. In order to ensure randomness in the initial distribution data gathering has to start after a certain simulation time. A mobile node starts simulation by waiting at one location for a specified pause time. After this time is over, it randomly selects the next destination in the simulation area. It also chooses a random speed uniformly distributed between a maximum and minimum speed and travels with a speed v whose value is uniformly chosen in the interval $(0, v_{max})$. Then the mobile node moves towards its selected destination at the selected speed. After reaching its destination, the mobile node again waits for the specified pause time before choosing a new way point and speed.

4.2. Movement Model

In the simulation environment the nodes move according to our selected random waypoint mobility model.We have generated the movement scenario files using the setdest program which comes with the NS-2 distribution. The total duration of our each simulation run is 900 seconds. We have varied our simulation with movement patterns for six different node speed: 5m/s, 10m/s, 15m/s, 20m/s, 25m/s, 30m/s. We have performed our experiment with two different numbers of source nodes: 30 source nodes and 40 source nodes. As slight changes in the movement pattern will have significant effect on the protocol performance, we have generated scenario files with 60 different movement patterns, 10 for each value of node speed. In order to compare the performance of the protocols based on the identical scenario both the protocols were run with these 60 different movement patterns.

4.3. Communication Model

In our simulation environment the MANET nodes use constant bit rate (CBR) traffic sources. We have used the *cbrgen* traffic-scenario generator tool available in NS2 to generate the CBR traffic connections between the nodes. Data packets transmitted are of 512 bytes. Data packets are sent at the rate of 5 packets/second.We have used two different communication patterns corresponding to 30 and 40 sources. The complete list of simulation parameters is shown in Table 1.

Table 1. Simulation Parameters.

Parameter	Value
Protocols	DSDV, DSR
Number of mobile nodes	70
Number of fixed nodes	10
Number of sources	30,40
Transmission range	250 m
Simulation time	900 s
Topology size	900 m X 600 m
Source type	Constant bit rate
Packet rate	5 packets/sec
Packet size	512 bytes
Pause time	100 seconds
Node speed	5m/s, 10m/s, 15m/s, 20m/s, 25m/s, 30m/s
Mobility model	Random way point

4.4. Hybrid Scenario

We have used a rectangular simulation area of 900 m x 600 m. The choice of rectangular area instead of square area was made in order to ensure longer routes between nodes. In our simulation we have used two ray ground propagation model. Our mixed scenario consists of a wireless and a wired domain. The simulation was performed with 70 wireless nodes and 10 wired nodes. For our hybrid networking environment we have a base station located at the centre (450,300) of the simulation area. The base station acts as a gateway between the wireless and wired domains. For our mixed simulation scenario we have turned on hierarchical routing in order to route packets between the wired and the wireless domains. The domains and clusters are defined by using the hierarchical topology structure. As the base station nodes act as gateways between the wired and wireless domains, they need to have their wired routing on. In the simulation setup we have done this by setting the node-config option –wiredRouting on. After the configuration of the base station, the wireless nodes are reconfigured by turning their wiredRouting off.

5. PERFORMANCE METRICS

We have primarily selected the following three performance metrics in order to study the performance comparison of DSDV and DSR.

Packet delivery fraction: This is defined as the ratio between the number of delivered packets and those generated by the constant bit rate (CBR) traffic sources.

Average end-to-end delay: This metric includes all possible delays caused by buffering at the time of the route discovery, queuing delay due to waiting at the interface queue, retransmission delays at MAC, propagation and transfer times. This is basically defined as the ratio between the summation of the time difference between the packet received time and the packet sent time and the summation of data packets received by all nodes.

Normalized routing load: This is defined as the number of routing packets transmitted per data packet delivered at the destination. Each hop-wise transmission of a routing packet is counted as one transmission.

5.1. Packet Delivery Fraction (PDF) Comparison

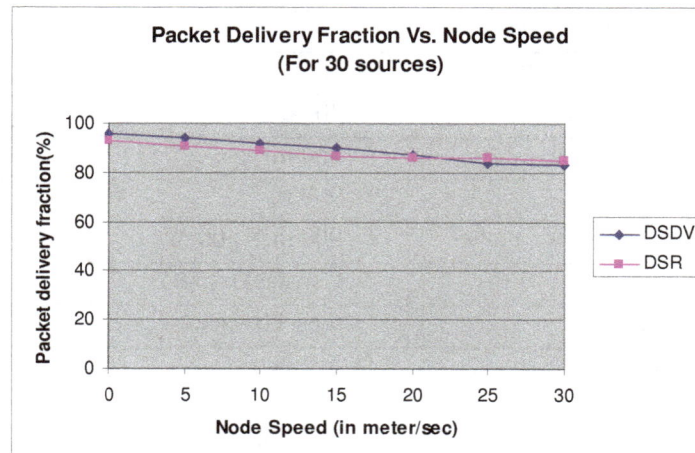

Figure 2. Packet Delivery Fraction Vs. Node Speed for 30 sources

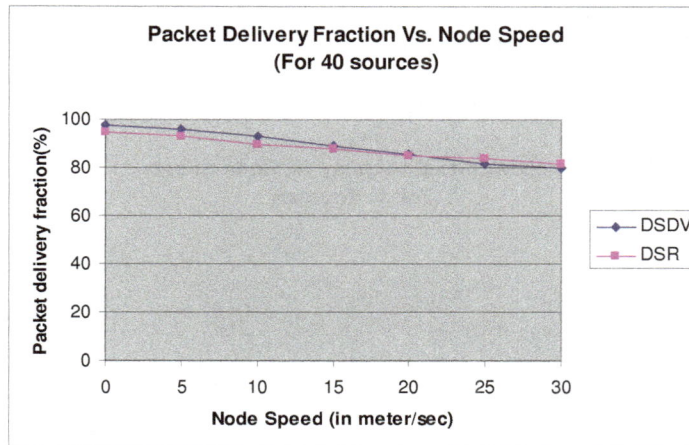

Figure 3. Packet Delivery Fraction Vs. Node Speed for 40 sources

From Fig. 2 and Fig. 3 we observe the difference in the packet delivery performances of DSDV and DSR from our simulation experiments. We have measured the packet delivery fraction of these two protocols by varying the node speed with respect to 30 and 40 numbers of sources. From the graphs we see that DSDV shows better packet delivery performance than DSR at lower node speed. This happens due to the fact that, at lower node speed, the network remains relatively stable and once a route is established, it continues to be available for a longer period of time. Due to the proactive nature of DSDV, routing information exchanges take place regularly between the nodes and each node maintains routing information to every destination all the time. Consequently, most of the packets can be delivered smoothly without having to wait for the path setup time. This results in better packet delivery performance of DSDV. On the contrary, DSR, being a source routing protocol, a significant time is required for initial path setup. During this time, no packets can be delivered to the destination due to unavailability of routes. This results in lower packet delivery fraction of DSR in comparison to DSDV.

With higher node speed, the network topology becomes highly dynamic and link breaks become more frequent. The unavailability of routes causes the nodes to show deterioration in the packet delivery performance for both DSDV and DSR. The periodic nature of operation of DSDV makes it less adaptive to these frequent changes. It requires greater number of full dumps to be exchanged between the nodes in order to maintain up-to-date routing information at the nodes. This huge volume of control traffic occupies a significant part of the channel bandwidth and lesser channel capacity remains available for the data traffic which results in reduced packet delivery fraction of DSDV at higher node speed.

DSR on the contrary, is more adaptive to the frequently changing scenario due to its on-demand nature of functioning. DSR maintains multiple routes in the cache. Thus, even if a link is broken due to higher node speed, alternative routes can be obtained from the cache. This reduces the number of dropped packets and results in better packet delivery performance of DSR.

It can also be noticed from the figures that as the number of sources is increased, initially when the network topology remains relatively stable at lower node speed, the packet delivery fraction also gets increased. This happens due to the fact that with lesser number of sources, the channel capacity is not fully utilized. Hence, increasing the number of sources also improves the packet delivery ratio. However, when the node speed is increased more along with greater number of sources, finding the route requires greater amount of routing traffic. This leads to reduced availability of the channel bandwidth for data transmission and more congestion which ultimately reduces the packet delivery ratio.

5.2. Average End-to-End Delay Comparison

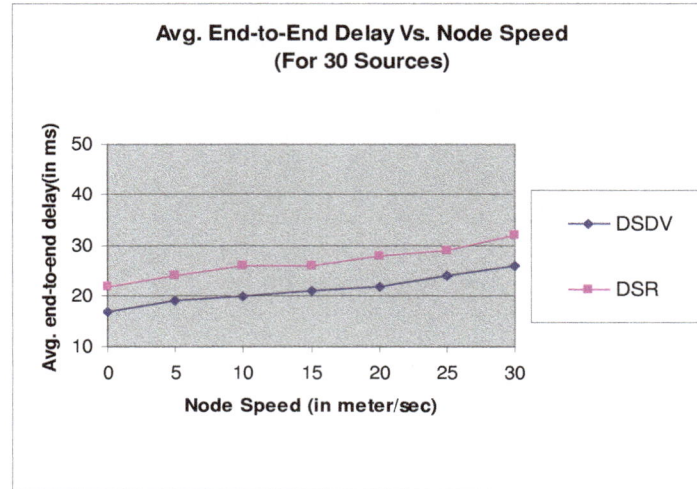

Figure 4. Average End to End Delay Vs. Node Speed for 30 sources

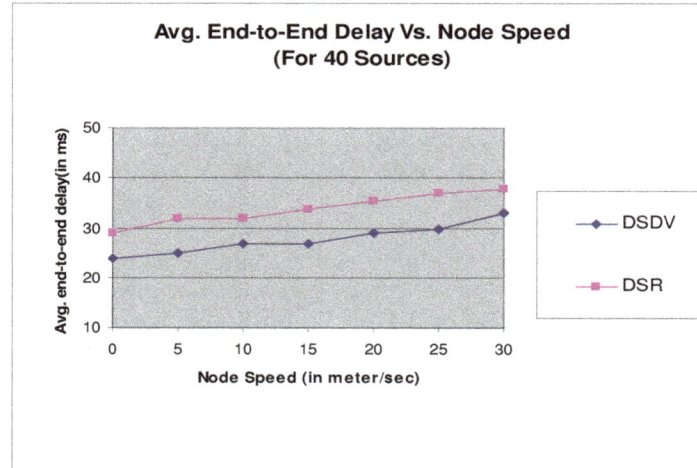

Figure 5. Average End to End Delay Vs. Node Speed for 40 sources

From Fig. 4 and Fig. 5 we can observe the fact that DSDV has less average end to end delay in comparison to DSR. DSDV is a proactive routing protocol. In DSDV, nodes periodically exchange routing tables between them in order to maintain up-to-date routing information to all destinations. Due to this regular route optimization, nodes have access to fresher and shorter routes to the destinations all the time. Hence, whenever a source node wants to send a packet to a destination node, with the already available routing information it can do so without wasting any time for path setup. This instant availability of fresher and shorter routes thus results in less average end-to-end delay in the delivery of data packets in case of DSDV.

DSR, on the contrary, is a reactive source routing protocol and routing information exchanges do not take place regularly. Instead, if a node in DSR wants to send a packet to a destination node, it has to first find the route to the destination in an on demand fashion. This route discovery latency is a part of the total delay. DSR being a source routing protocol, the initial path set up time is significantly higher as during the route discovery process, every intermediate node needs to extract the information before forwarding the data packet. Moreover in DSR, the source needs to wait for all the replies sent against every request reaching the destination. This increases the delay.

From the figures it is evident that the average end-to-end delay becomes more with higher node speed and greater number of sources for both the protocols. Frequent changes in the network topology due to increasing node speed results in greater number of link breaks. This together with the greater number of sources requires DSR to invoke the route discovery process more frequently in order to find new routes. The frequent invocation of the route discovery creates huge amount of control traffic. The data traffic to be delivered also becomes more with greater number of sources. This results in more collisions, further retransmissions and higher congestion in the network. Consequently, the route discovery latency increases due to the constrained channel. This in turn increases the average end-to-end delay. In addition to that, due to the higher priority of the control packets, the data packets need to spend more time in the queue waiting for the huge volume of control packets to be delivered. This also increases the end-to-end delay in delivering the data packets. In case of DSDV, due to higher speed of the nodes and frequent link breaks, routes become unavailable and nodes need to wait till the next routing information exchanges for new routes. Thus the delay increases depending upon the duration of the interval between the successive routing information exchanges.

5.3. Normalized Routing Load Comparison

From Fig. 6 and Fig. 7 we note that initially at lower node speed, DSR has greater normalized routing load. This is attributed to the fact that DSR being a source routing protocol, with every packet the entire routing information is embedded. In addition to that, in response to a route discovery, replies come from many intermediate nodes. This increases the total control traffic. In case of DSDV, initially, at lower node speed, the network topology remains relatively stable. Hence, nodes need to exchange only incremental dumps rather than full dumps. This results in lesser overhead of DSDV.

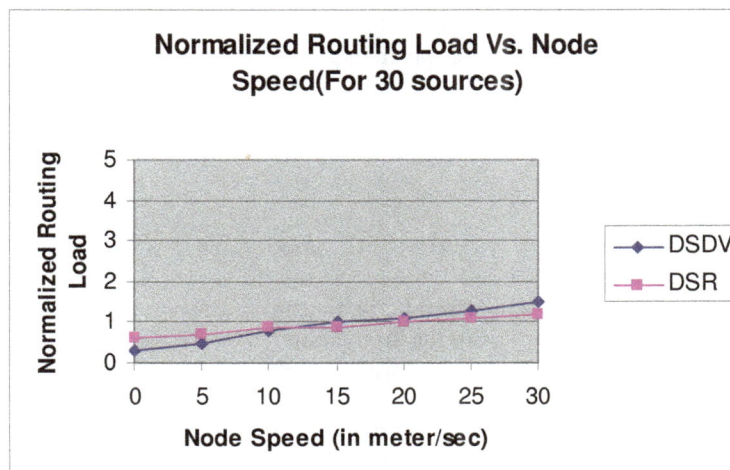

Figure 6. Normalized Routing Load Vs. Node Speed for 30 sources

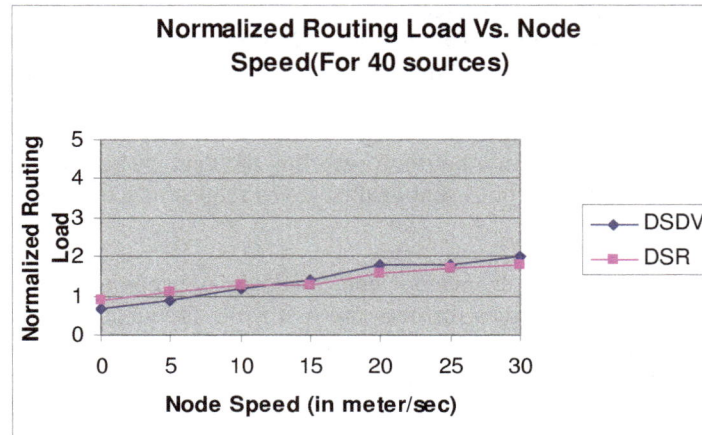

Figure 7. Normalized Routing Load Vs. Node Speed for 40 sources

Both DSDV and DSR suffer from increased normalized routing load with higher node speed and greater number of sources. In case of DSR, with increasing node speed, the route discoveries need to be invoked more often due to increase in the number of broken links. Furthermore, as DSR does not use route optimization until the route is broken and continues using longer and older routes, the chances of link breaks also increase. This further adds to the number of route discoveries which ultimately results in huge control traffic and subsequently higher normalized routing load. Greater number of sources also causes frequent invocation of the route discovery which significantly increases the volume of control overhead. Higher volume of data and control traffic creates congestion in the network. This results in further collisions, more retransmissions and newer route discoveries and further adds up to the already increased control overhead which ultimately results in higher normalized routing load.

With higher node speed, the network topology experiences frequent and high volume of changes. DSDV, due to its proactive nature of operation, is less adaptive to this highly dynamic scenario. Therefore, nodes need to exchange full dumps in order to maintain up-to-date routing information. This causes greater routing overhead for DSDV. In comparison, DSR uses aggressive caching strategy and the hit ratio is quite high. As a consequence, in highly dynamic scenario, even if a link breaks, DSR can resort to an alternate link already available in the cache. Thus the route discovery process can be postponed until all the routes in the cache fail. This reduces the frequency of route discovery, which ultimately results in less routing overhead of DSR.

6. CONCLUSIONS

In this paper we have carried out a detailed ns2 based simulation to study and analyze the performance differentials of DSDV and DSR in the hybrid scenario under varying node speed with different number of sources. Our work is the first in an attempt to compare these protocols in hybrid networking environment. From the simulation results we see that at lower node speed, DSDV shows better packet delivery performance than DSR mainly due to the instant availability of fresher and newer routes all the time. On the other hand, with higher node speed, DSDV shows more deterioration in the packet delivery performance than DSR mainly due to its less adaptability to the highly dynamic network topology. DSR's better performance is attributed to its ability to maintain multiple routes per destination and its use of aggressive caching strategy. In terms of the average end-to-end delay, DSDV outperforms DSR. The poor performance of DSR in terms of average end-to-end delay is primarily due to its source routing nature and its inability to expire the stale routes. Both the approaches suffer form greater average end-to-end delay when we increase the speed of the nodes and the numbers of sources. At higher node speed we observe that DSR shows lower routing load in comparison to DSDV. DSR applies aggressive caching technique and maintains multiple routes to the same destination. Hence, in highly dynamic scenario, even if a link is unavailable due to link break, DSR can resort to an alternate link already available in the cache. This results in reduced frequency of route discovery which ultimately reduces the routing overhead of DSR. On the other hand, at lower node speed, the network topology remains relatively stable. Hence, in DSDV, nodes need to exchange only incremental dumps rather than full dumps. This results in lesser overhead of DSDV. Thus we can conclude that if routing delay is of little concern, then DSR shows better performance at higher mobility in terms of packet delivery fraction and normalized routing load in hybrid networking scenario. Under less stressful scenario, however, DSDV outperforms DSR in terms of all the three metrics.

REFERENCES

[1]   Dow, C. R, (March 2005) A Study of Recent Research Trends and Experimental Guidelines in Mobile Ad-Hoc Networks, In: Proceedings of 19th International Conference on Ad-vanced Information Networking and Applications, IEEE, Vol. 1, pp. 72-77.

[2]   Freisleben, B., Jansen, R, (1997) Analysis of Routing Protocols for Ad hoc Networks of Mobile Computers, In: Proceedings of the 15th IASTED International Conference on Applied Informatics, IASTED-ACTA Press, pp. 133-136, Innsbruck, Austria.

[3]   Royer, E. M. , Toh, C. K, (April 1999) A Review of Current Routing Protocols for Ad hoc Mobile Wireless Networks, IEEE Personal Communications Magazine, pp. 46-55.

[4]   Anastasi, G., Borgia, E., Conti, M., Gregori, E, (2003) IEEE 802.11 Ad-hoc Networks: Protocols, Performance and Open Issues, Ad hoc Networking. IEEE Press Wiley, New York.

[5]   Arun Kumar, B. R., Reddy, Lokanatha C., Hiremath, Prakash S, (January-June 2008) A Survey of Mobile Ad hoc Network Routing Protocols, Journal of Intelligent System Research.

[6]   Rappaport, T. S, (1996) Wireless Communications, Principles & Practices. Prentice Hall.

[7]   Vaidya, Nitin H, (2004) Mobile Ad Hoc Networks: Routing, MAC and Transport Issues, Tutorial presented at INFOCOM 2004 (IEEE International Conference on Computer Communication), University of Illinois at Urbana-Champaign.

[8]   Arun Kumar, B. R., Reddy, Lokanatha C., Hiremath, Prakash S, (2008) Mobile Ad hoc Networks: Issues, Research Trends and Experiments, International Engineering and Technology (IETECH) Journal of Communication Techniques, Vol. 2, No. 2.

[9]   Toh, C. K, (2002) Ad-Hoc Mobile Wireless Networks. Prentice Hall.

[10]   Tanenbaum, Andrew S, (2002) Computer Networks. Fourth Edition. Prentice Hall.

[11]   Corson, S. , Macker, J, (January 1999) Mobile Ad hoc Networking (MANET): Routing Protocol Performance Issues and Evaluation Considerations, IETF MANET Working Group RFC-2501.

[12]   Murthy, C. S. R., Manoj, B. S, (2004) Ad Hoc Wireless Networks: Architecture and Protocols, Prentice Hall Communications, Engineering and Emerging Technologies Series, New Jersey.

[13]   Comer, D, (2000) Internetworking with TCP/IP, Volume 1, Prentice Hall.

[14]   Blum, Jermy I., Eskandarian, Azim , Ho_man, Lance J, (Dec. 2004) Challenges of inter-vehicle Ad hoc Networks, IEEE transactions on Intelligent Transportation Systems. Vol. 5, No. 4.

[15]   Das, S. R., Castaeda, R., Yan, J, (2000) Simulation-based Performance Evaluation of Routing Protocols for Mobile Ad hoc Networks, Mobile Networks and Applications, Vol. 5, pp. 179-189.

[16]   Das, Samir R., Perkins, Charles E., Royer, Elizabeth M, (March 2000) Performance Comparison of two On-demand Routing Protocols for Ad hoc Networks, In: Proceedings of the IEEE Conference on Computer Communications (INFOCOM), Tel Aviv, Israel, pp. 3-12.

[17]   Johansson, P., Larsson, T., Hedman, N., Mielczarek, B, (August 1999) Routing Protocols for Mobile Ad-hoc Networks - A Comparative Performance Analysis, In: Proceedings of the 5th International Conference on Mobile Computing and Networking (ACM MOBICOM'99).

[18]   Park, Vincent D., Corson, M. Scott, (June 1998) A Performance Comparison of TORA and Ideal Link State Routing, In: Proceedings of IEEE Symposium on Computers and Communication '98.

[19]   Fall, K., Vardhan, K. Eds, (1999) Ns notes and documentation, available from, http://www.mash.cd.berkeley.edu/ns/.

[20]   Network Simulator-2 (NS2), http://www.isi.edu/nsnam/ns

[21]   The CMU Monarch Project: The CMU Monarch Projects Wireless and Mobility Extensions to ns, http://www.monarch.cs.cmu.edu (1998).

[22]   Altman, E., Jimenez, T, (2003) NS Simulator for Beginners, Lecture notes. Univ. de Los Andes, Merida, Venezuela and ESSI. Sophia-Antipolis, France.

[23]   IEEE Computer Society LAN MAN Standards Committee, (1997) Wireless LAN Medium Access Control (MAC) and Physical Layer (PHY) Specifications, IEEE Std 802.11-1997. The Institute of Electrical and Electronics Engineers, New York.

8

Analysis of the Paging Costs in Location Management for Mobile Communications Networks Leveraging the Diffusion Constant

E. Martin and R. Bajcsy

Department of Electrical Engineering and Computer Science
University of California, Berkeley
California, USA

emartin@eecs.berkeley.edu

Abstract

The recent growth of users in mobile communications networks has shown the need to efficiently manage the signalling burden related to Mobility Management. Within this field, the optimisation of Location Management costs has become a key research topic. In this sense, the development of accurate mobility models to describe the user's behaviour is still a pending task. In this article, having characterized the diffusion constant for distinct types of movements, the evaluation of the paging costs for different motion processes through the diffusion constant is reviewed, obtaining useful guidelines for their ranking in terms of signalling costs. As a consequence of this analysis, a novel parameter is introduced to account for how searchable a mobile user is, and whose inverse approximates the paging costs of a wide variety of motion processes.

Keywords

Paging Costs, Diffusion Constant, Mobility Management, Location Management, Mobile Communications Networks

1. Introduction

Location Management is a topic of increasing importance due to the rise in the number of users in mobile communications networks [1-6]. Most of the recent research in this field has focused on the evaluation of the signalling costs involved in both location update and paging [7-10]. Common techniques to assess Location Management signalling costs make use of time-varying probability distributions on the mobile user's location, derived either from motion models or approximated by means of empirical data [11-17]. This strategy is especially suitable when the mobile terminal changes location according to stochastic processes. Considering for example an isotropic Brownian motion process with drift, the Gaussian probability density function for the location of the mobile can be applied [18], and the location probability can be obtained through the integration of the density function over the region of interest. For the particular analysis of the paging costs, an approach based on location probability distributions could be the information theoretical one, by means of relating the entropy of the distribution with the cost of paging, as the mean of the ordered distribution corresponds to the minimum paging costs [19]. However, this strategy faces the difficulty that the entropy and the mean number of locations paged can present very large variations.

In this article, we examine the evaluation of the paging costs for different motion processes leveraging the diffusion constant. As a consequence of this analysis, we introduce a novel parameter to account for how searchable a mobile user is, and whose inverse approximates the paging costs of a wide variety of movements. The rest of this article is organized as follows: in Section 2 we analyze the behaviour of the diffusion constant for one-dimensional, two-dimensional and planned movements. In Section 3, leveraging the diffusion constant to calculate the paging costs of different motion processes, we perform comparisons between them in order to obtain guidelines on their costs' ranking for different time intervals and mobility parameters. In Section 4 we introduce a novel parameter with the name Searchability Index, useful for the approximation of the paging costs. Conclusions are drawn in Section 5.

2. STUDY OF THE DIFFUSION CONSTANT FOR DIFFERENT TYPES OF MOVEMENTS

Considering isotropic Brownian motion process with drift, assuming movements of the user to the right or to the left with probability p or q respectively, the Gaussian probability density function for the location of the mobile can be applied [20], and two parameters with special importance in this type of processes can be leveraged for a deeper characterization of the motions: 1) the mean drift velocity v, defined as:

$$v = (p - q)\frac{\Delta x}{\Delta t} \tag{1}$$

and 2) the diffusion constant Dif :

$$Dif = \left((1 - p)p + (1 - q)q + 2pq\right)\frac{(\Delta x)^2}{\Delta t} \tag{2}$$

where Δx and Δt represent the space and time steps. Next we study the behaviour of the diffusion constant for different types of movements.

2.1. Study of the diffusion constant for one-dimensional movements.

A typical random movement where the mobile user can change direction randomly at each point can be analysed through the consideration of a Brownian motion process, with p and q taking values according to the type of movement. For example, for a one-dimensional movement along a straight line, if it is considered equally likely to move forward or backwards, then $p=q$ and $1-p-q$ is the probability to stay in the same point. However, p and q can take different values, thus reflecting certain preference forward or backward. In Figure 1 the variability of the diffusion constant with the mean velocity is shown for different values of p and q.

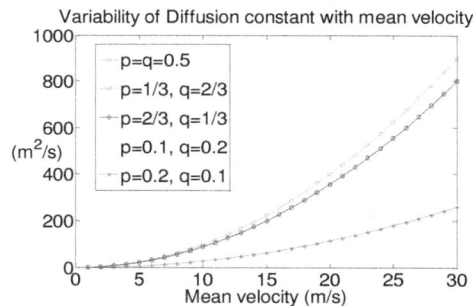

Figure 1. Variability of diffusion constant with mean velocity for one-dimensional random movement.

In Figure 1, it can be seen that the diffusion constant increases parabolically with the mean velocity of the user. At the same time, the diffusion constant value does not vary if the p and q values are exchanged. Therefore, the diffusion constant reflects the mobility of the user, regardless of the direction of the movement. It is interesting to note that the sensors embedded in current state-of-the-art smart phones can be leveraged to obtain very precise information about the mobility of the user [12-17], which can be very helpful to manage the location of the mobile communications networks' users. Another clear consequence inferred from Figure 1 is the fact that as the p or q values increase, so does the diffusion constant, but for combinations of p and q which add up the same quantity, the diffusion constant will be higher for those values of p and q that enclose the larger area in a theoretical rectangle of dimensions $p \cdot q$.

2.2. Study of the diffusion constant for two-dimensional movements.

A two-dimensional movement where the user moves randomly towards any of four typical possible directions (North, South, East, West) at each point, can be split into two one-dimensional movements with the same mobility characteristics, but half probabilities. That is, if the movement to any of the four possible directions is equally likely, p and q will be 0.25 for every one-dimensional version. In Figure 2, the influence of the addition of one more dimension for the movement in the diffusion constant is analysed.

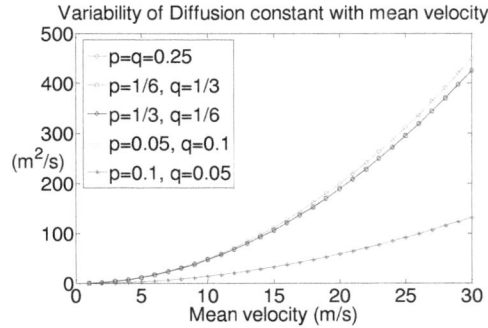

Figure 2. Variability of diffusion constant with mean velocity for two-dimensional random movement.

The diffusion constant for every one-dimensional version decreases approximately by half in comparison with the previous case, but the behaviour with the mean velocity and the p and q values is the same.

2.3. Study of the diffusion constant for planned movements.

The previous model can also be applied to situations such as those planned movements with constant speed and no random changes of direction, where theoretically the diffusion constant would be null. Carefully studying these movements [21, 22], taking into account the standard deviation in arrival time dev, the diffusion constant can be expressed as follows:

$$Dif = \frac{2 \cdot (V \cdot dev)^2}{Duration} \tag{3}$$

Where $Duration$ is the average trip duration and V represents the average velocity. For example, for a planned trip of 2 hours, considering standard deviations in arrival time ranging from 1 minute to 10 minutes, the values that can be obtained for the diffusion constant are displayed in Figure 3.

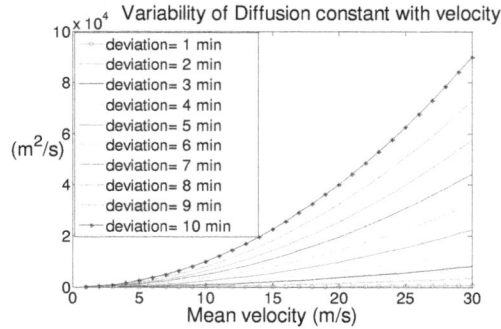

Figure 3. Variability of diffusion constant with mean velocity for planned movement.

It can be noticed the way in which the diffusion constant rises with the deviation in arrival times, and again, with the velocity. The increase of the diffusion constant with the value of the deviation is gradual. For higher velocities, such as those for high-speed trains or aircrafts, the values taken can be observed in Figure 4.

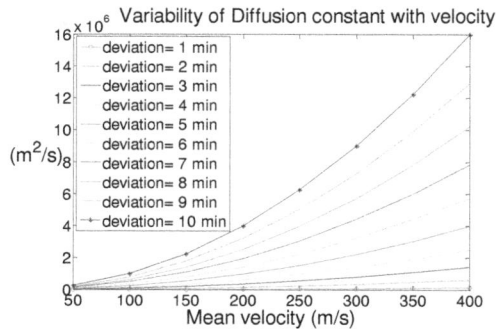

Figure 4. Variability of diffusion constant with mean velocity for planned movement and high velocity.

3. APPLICATION OF THE DIFFUSION CONSTANT FOR THE CALCULATION OF THE PAGING COSTS.

An approximation for the minimum of the paging costs by means of the diffusion constant is obtained in [23]: $(Dif \cdot t)^{\frac{n}{2}}$, where t is the time elapsed since the last interaction mobile user-network, and n is the movement dimension. In Figure 5, the variability of the paging costs for Brownian motions of various dimensions and for different call arrival rates is shown:

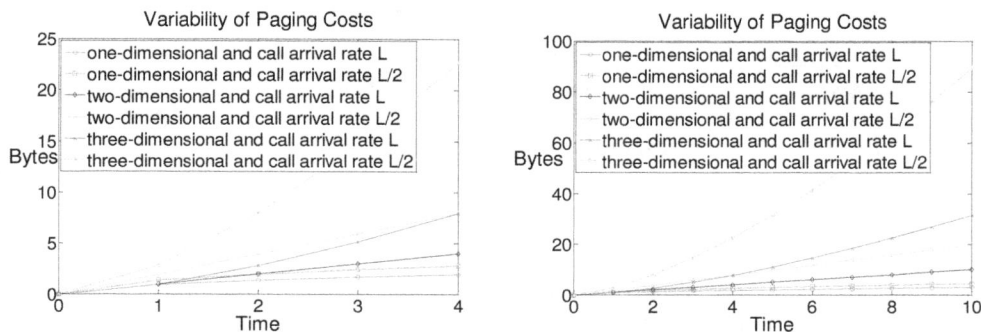

Figure 5. Variability of paging costs for different movement dimensions and call arrival rates, for short and long time intervals.

For simplicity purposes, the diffusion constant for the movements in Figure 5 has been taken as unity, and the time units are considered in terms of call inter-arrival periods. Comparing motion processes of different characteristics, for very short time intervals elapsed from the last interaction mobile user-network (typically a number of time units coincident with the value of the lower call-to-mobility ratio of the movements being compared, or one time unit in case that lower ratio is below unity), the paging costs for each particular call are very similar for all the different movement dimensions and call arrival rates, but the three-dimensional one with the lowest call arrival rate tends to outstand from the rest for presenting the fastest increase in time of the paging costs, with a parabolic evolution. The next movement with the highest paging costs is the two-dimensional one with the lowest call arrival rate, followed by the one-dimensional one with the lowest call arrival rate. Consequently, at these early time intervals, although the higher the movement dimension, the larger the paging costs for processes with call-to-mobility ratios below unity (opposite behaviour for those processes with call-to-mobility ratios above unity), it is clearly noticed the key role played by the call arrival rate to classify the motion processes in terms of their paging costs: the lower the call arrival rate, the larger the paging costs for each particular call.

Still at very early stages, but for longer time intervals than in the previous considerations, the three-dimensional movement with high call arrival rate starts increasing its paging costs above those movements with lower dimensions and lower call arrival rates.

Comparing two motion processes with different call arrival rate, and no restrictions for their particular diffusion constants, we define a threshold value we name Dimensionality Threshold (D_{TH}), measured in time units, and obtained by means of the following expression we derive in the Appendix I:

$$D_{TH} = \frac{(CMR_h)^3}{(CMR_l)^2} \tag{4}$$

Where CMR_h and CMR_l are the call-to-mobility ratios (high and low respectively) of the two types of processes being compared through their call arrival rates. The definition of this threshold is:

> Dimensionality Threshold: Length in time units over which the call arrival rate exchanges its role with the movement dimension as the key parameter to classify motion processes of different characteristics according to the value of their paging costs for each particular call.

As the time from the last interaction exceeds the Dimensionality Threshold, the three-dimensional movement with low call arrival rate clearly becomes the one with the largest paging cost, followed by the three-dimensional one with high call arrival rates, next the two-dimensional ones accordingly to their call arrival rates, and eventually the one-dimensional ones, again with larger paging costs for lower call arrival rates.

In Figure 6, the variability of the paging costs for Brownian motions of different dimensions and for different diffusion constants is shown:

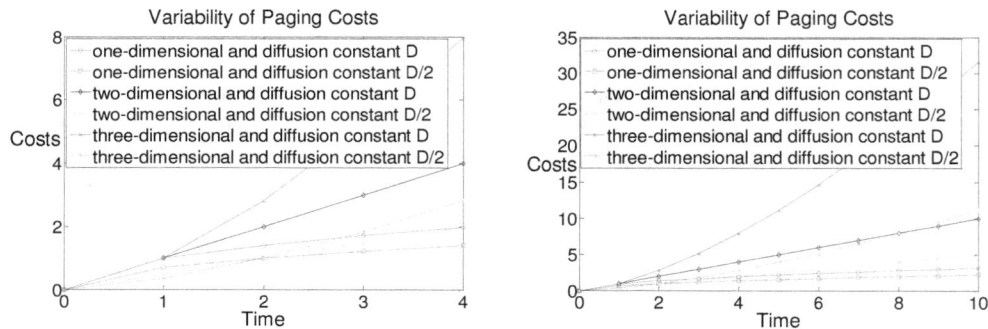

Figure 6. Variability of paging costs for different movement dimensions and diffusion constant values, for short and long time intervals.

For simplicity purposes, the call arrival rates for the motion processes in Figure 6 have been taken as unity. In order to perform the comparison related to the different movement characteristics, within the first of the shorter movement periods from the last interaction mobile user-network (although the same behaviour can be noticed within the first of the longer movement periods), the dimensionality of the movement has little importance to rank the paging costs. At this stage, the value of the diffusion constant rules the behaviour of the costs, and the higher the diffusion constant, the larger the paging costs. Actually, for these time intervals, in opposition to the behaviour of processes with call-to-mobility ratios below 1, for call-to-mobility ratios above 1 it can be observed that the lower the movement dimension, the larger the paging costs. It is interesting to think about an explanation for this behaviour. The reasoning we propose is based on the fact that assuming a fixed amount of resources for a particular location area regardless of the movement dimensionality, the higher the movement dimension for diffusivities with a frequency lower than the call arrival rate, within the first movement period, the higher the mobile's specifiability in terms of amount of resources needed to track him down, therefore the lower that amount of resources, thus diminishing the paging costs for each particular call. Once the first movement period is passed, the higher the movement dimension, the higher the uncertainty of the next location of the mobile, thus the more dramatically the mobile's specifiability declines, therefore making the paging costs for each particular call larger. The basic idea of the previous statements is illustrated in a simplified way with Figure 7:

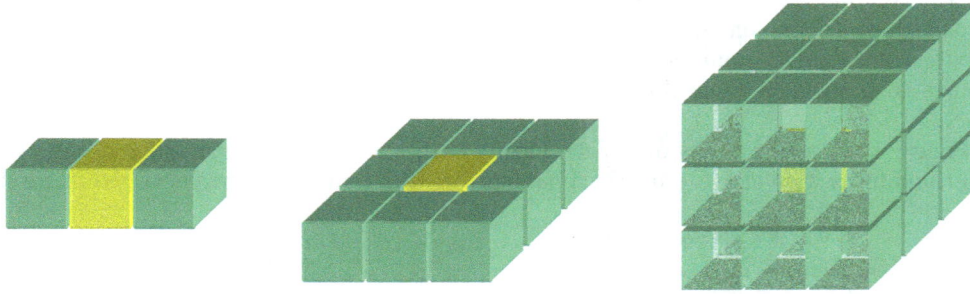

Figure 7. Variation of the mobile's specifiability with the movement dimension.

Assuming that the number of cells per location area increases with the movement dimension, when the movement frequency is lower than the paging frequency, within the first movement period, the amount of resources being used (R_s) specifically to track down the mobile in the deployments represented in Figure 7 is:

$$R_s\big|_{t<T} = \left(\frac{1}{3^n}\right) \cdot R \qquad (5)$$

Where R is the total amount of resources used for each location area, n is the movement dimension, t represents the time elapsed since the last interaction mobile-network, and T is the movement period. As expected, R_s falls when n rises. On the other hand, once the first movement period is passed, the uncertainty for the whereabouts of the mobile increases with the movement dimension, and thus the amount of resources needed to track him down increases with the movement dimension, as shown by the new expression taken by R_s:

$$R_s\big|_{t>T} = \left(\frac{3^n - 1}{3^n}\right) \cdot R \qquad (6)$$

Graphically, assuming $R=1$ for simplicity purposes, the behaviour of R_s with the movement dimension is presented in Figure 8:

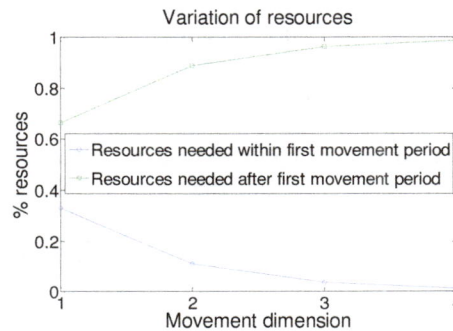

Figure 8. Variation of resources needed to track down a mobile with diffusivity frequency lower than call arrival rate, before and after the first movement period.

As observed in Figure 8, the amount of resources needed varies with the movement dimension, in a way dependant on the time interval considered. It would be interesting to extend the previous reasoning beyond mobile communications networks users, and consider movements of entities in more than three dimensions. However we will not tackle that discussion in this article.

In Figure 9, the actual behaviour of the paging costs for each particular call for processes with call-to-mobility ratio above 1 and for time intervals within the first movement period are shown:

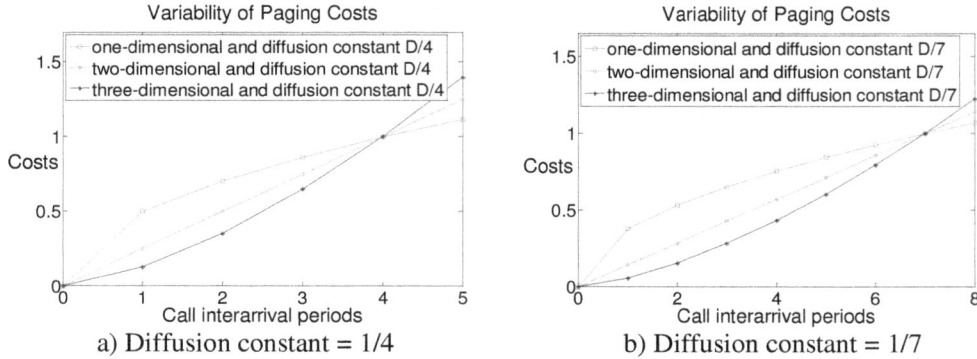

a) Diffusion constant = 1/4   b) Diffusion constant = 1/7

Figure 9. Variability of paging costs for different dimensions and diffusion constants providing a movement rate with longer period than the call inter-arrival one.

Continuing with the previous comparison between motion processes of different dimensions and diffusion constants, in between the first and the second movement period of the process with the highest call-to-mobility ratio of those being compared, although the diffusion constant remains as the main factor for the value of the paging costs, it starts becoming noticeable that the higher the movement dimension, the faster the paging costs increase with time.

For longer time intervals, the movement dimension keeps gaining importance regarding the value of the paging costs, and the diffusion constant continues loosing it, especially for one and two-dimensional movements. Making use of the Dimensionality Threshold again (in this case its definition will simply substitute the call arrival rate by the diffusion constant, while the mathematical formula remains the same), once the time elapsed from the last interaction user-network passes it, the paging costs of the motion processes are ranked firstly by their movement dimension (costs rise with it), and within each dimension, the higher the diffusion constant, the larger the paging costs become.

4. SEARCHABILITY INDEX

Given the importance of location information and context awareness in mobile computing [24-28], next we leverage the results previously obtained for the behaviour of the paging costs with different process dimensions and call-to-mobility ratios, and we introduce a novel parameter we call Searchability Index (*SI*), which will provide us with a good indication of how searchable a mobile user is. In other words, *SI* will provide us with an indication of the granularity in the location information that the system has about the user, which will be very helpful for the network operators to optimize the Location Management procedures [2-6]. We have inferred the mathematical definition of the *SI* from the results obtained analyzing multiple cases:

$$SI = Tc/(Pc)^{\left[Int\left(\frac{Dif}{\lambda}\cdot t\right)\right]} \tag{7}$$

With t representing the time elapsed from the last interaction mobile user-network, the exponent of the denominator being the integer part of that time normalized to the value of the call-to-mobility ratio, Pc symbolizing the number of possible new cells where the mobile can enter every time it moves, and Tc representing the total number of cells in its area of immediate influence. These two last quantities depend on the movement dimension. Defining a parameter we call "cluster one" (C_1) as the numerical size of the area being administered in the one-dimensional case (for example, number of cells per location area), Pc and Tc take the following expressions:

$$Pc = (C_1)^n - 1 \tag{8}$$

$$Tc = (C_1)^n \tag{9}$$

With n representing the movement dimension. For simplicity purposes, we have chosen 3 as the reference value for C_1.

The meaning of the Searchability Index can be inferred from the fact that the larger its value, the more predictable the whereabouts of the mobile user are. Its values would range between 0 and ∞. The former for extremely unpredictable locations, and the latter for absolute certainty in the whereabouts of the mobile. The Searchability Index does not have units, and is an illustrative figure whose inverse approximates the value of the paging costs for each particular call for movements of different characteristics. In order to obtain a better approximation for the value of the costs over selected normalized time spans, we split the time into intervals of length unity by means of the integer part function.

The behaviour of the inverse of the Searchability Index is shown graphically in Figure 10:

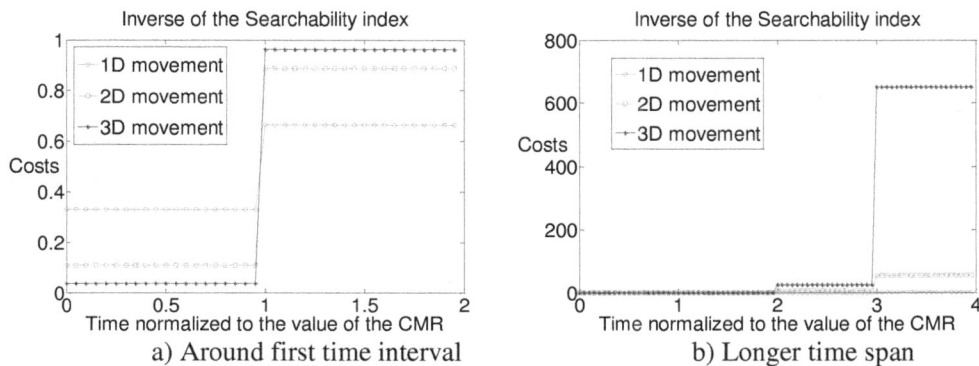

a) Around first time interval   b) Longer time span

Figure 10. Evolution of the inverse of the Searchability Index.

However, through the study of series of average values of the paging costs and series of values of the inverse of the Searchability Index in the time intervals considered, a further refinement in the definition of the Searchability Index is possible in order to obtain a better approximation between both terms, by means of taking into account some more influences of parameters involved in the motion process. In this sense, we derive the following expression:

$$SI = \cfrac{Tc \cdot \left\{ \sqrt{(C_1)^{n \cdot \left[Int\left(\frac{Dif}{\lambda} \cdot t\right) - 1 \right]}} \right\}^{u\left[Int\left(\frac{Dif}{\lambda} \cdot t\right) \right]}}{\left(\cfrac{Pc}{n}\right)^{\left[Int\left(\frac{Dif}{\lambda} \cdot t\right) + 1 \right]}} \tag{10}$$

With $u(t)$ in the exponent of the numerator, defined as follows:

$$u(t) = \begin{cases} 0 & t < 1 \\ 1 & t \geq 1 \end{cases} \tag{11}$$

And now, through the inverse of this refined Searchability Index, a better approximation for the value of the paging costs can be obtained. In Figure 11, the evolution of the inverse of this refined Searchability Index is shown for different normalized time lengths:

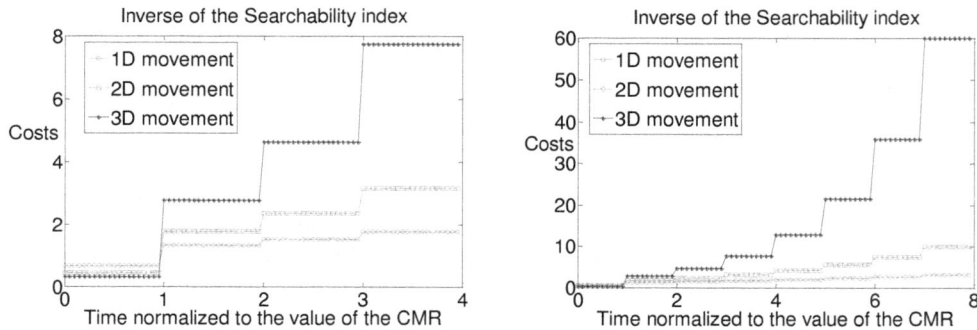

Figure 11. Evolution of the inverse of the refined Searchability Index, for short and long time intervals.

This Searchability Index accounts not only for the movement dimension and call-to-mobility ratio, but also for the network topology, in particular, the size of the clusters of cells. Its range of applicability through the formula we derived is: $C_1 > 2$, and length of the time in study below 10 normalized units, as otherwise, the results obtained would present higher values than the actual costs for the three-dimensional case. In Figure 12, a comparison between the actual paging costs and the inverse of the Searchability Index is shown for an interval of 4 time normalized units (average accuracies above 82%, with larger values for one- and two-dimensional movements):

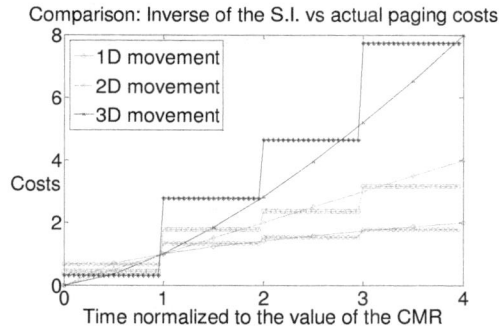

Figure 12. Comparison between the inverse of the Searchability Index and actual paging costs.

Additionally, Figure 12 shows that the inverse of the Searchability Index perfectly accounts for the exchange after the first normalized time unit between the movement dimensions in the ranking of more costly processes, when those processes have a call-to-mobility ratio above unity. The inverse of the Searchability Index also approximates correctly the paging costs for motion processes with call-to-mobility ratios below unity, for which the referred exchange between dimensions does not take place.

Having studied the behaviour of the inverse of the Searchability Index for different values of C_1 and time intervals, we can conclude that expectedly, the larger the value of C_1, the greater the resulting paging costs.

5. CONCLUSIONS.

The analysis of the diffusion constant shows that it accounts for the mobility of the user regardless of the direction of the movement. Applying the concept of the diffusion constant to obtain an approximation of the paging costs for Brownian movements of different dimensions, we reach the following conclusions: within the first units of the time elapsed from the last interaction mobile-network, although the higher the movement dimension, the larger the paging costs for processes with call-to-mobility ratios below unity (opposite behaviour for those processes with call-to-mobility ratios above unity), it is clearly noticed the key role played by the call arrival rate to rank the different movements; in fact, the lower the call arrival rate, the larger the paging costs for each particular call. Once the time elapsed from the last interaction mobile-network exceeds the Dimensionality Threshold, the movement dimension is the decisive parameter in the rise of the paging costs, and the call arrival rate takes a secondary role, especially for the one and two-dimensional movements. In particular, at these time intervals, the higher the dimensionality, the larger the paging costs, and for a same movement dimension, the lower the call arrival rate, the larger the paging costs for each call. An analogous study can be carried out in terms of the movement dimension and the diffusion constant, where the latter will show trends inverse to the call arrival rate.

In order to account for the behaviour of the paging costs of different motion processes, we have introduced a novel parameter called Searchability Index, whose meaning is based on the fact that the larger it becomes, the more predictable the whereabouts of the mobile user are, and its inverse approximates the value of the paging costs. The Searchability Index considers not only the dimensionality and call-to-mobility ratio of the process, but also the network topology, specifically, the size of the clusters of cells. It is important to notice that the inverse of the Searchability Index perfectly accounts for the exchange between the movement dimensions in the ranking of more costly movements after the first normalized time unit, for processes with call-to-mobility ratio above unity. The inverse of the Searchability Index is also consistent with motion processes whose call-to-mobility ratio is below unity and for which the referred exchange between dimensions does not take place.

5. REFERENCES

[1]   H. Vijayakumar, M. Ravichandran, "Efficient location management of mobile node in wireless mobile ad-hoc network", Proceedings of National Conference on Innovations in Emerging Technology, NCOIET'11, p 77-84, 2011

[2]   E. Martin, R. Bajcsy, "Savings in Location Management Costs Leveraging User Statistics", International Journal of Ubiquitous Computing, July, 2011.

[3]   E. Martin, R. Bajcsy, "Variability of Location Management Costs with Different Mobilities and Timer Periods to Update Locations", International Journal of Computer Networks & Communications, July, 2011.

[4]   E. Martin, "A graphical Study of the Timer Based Method for Location Management with the Blocking Probability", International Conference on Wireless Communications, Networking and Mobile Computing, 2011.

[5]   E. Martin, "Characterization of the Costs Provided by the Timer-based Method in Location Management", International Conference on Wireless Communications, Networking and Mobile Computing, 2011.

[6]   E. Martin, "New Algorithms to Obtain the Different Components of the Location Management Costs", International Conference on Wireless Communications, Networking and Mobile Computing, 2011.

[7]   J. R. Gállego, M. Canales, A. Hernández-Solana, A. Valdovinos, "Adaptive paging schemes for group calls in mobile broadband cellular systems", IEEE International Symposium on Personal, Indoor and Mobile Radio Communications, PIMRC, p 2444-2449, 2010

[8]   L. Jong-Hyouk, P. Sangheon, Y. Ilsun, C. Tai-Myoung, "Enabling a paging mechanism in network-based localized mobility management networks", Journal of Internet Technology, v 10, n 5, p 463-472, 2009

[9]   G. Ashish, G. Navankur, K. Prakhar, "A speed based adaptive algorithm for reducing paging cost in cellular networks", Proceedings - 2009 2nd IEEE International Conference on Computer Science and Information Technology, ICCSIT 2009, p 22-25, 2009.

[10]   V. Casares-Giner, P. García-Escalle, "A lookahead strategy for movement-based location update in wireless cellular networks", ITNG 2009 - 6th International Conference on Information Technology: New Generations, p 1171-1177, 2009.

[11]   S. Madhavapeddy, K. Basu, A. Roberts, "Adaptive paging algorithms for cellular systems", IEEE Vehicular Technology Conference, 1995, v 2, p. 976-980.

[12]   E. Martin, "Solving Training Issues in the Application of the Wavelet Transform to Precisely Analyze Human Body Acceleration Signals", IEEE International Conference on Bioinformatics and Biomedicine, 2010.

[13]   E. Martin, R. Bajcsy, "Analysis of the Effect of Cognitive Load on Gait with off-the-shelf Accelerometers", International Conference on Advanced Cognitive Technologies and Applications, 2011.

[14]   E. Martin, R. Bajcsy, "Considerations on Time Window Length for the Application of the Wavelet Transform to Analyze Human Body Accelerations", IEEE International Conference on Signal Processing Systems, 2011.

[15]   E. Martin, "Optimized Gait Analysis Leveraging Wavelet Transform Coefficients from Body Acceleration", International Conference on Bioinformatics and Biomedical Technology, 2011.

[16]   E. Martin, "Novel Method for Stride Length Estimation with Body Area Network Accelerometers", IEEE Conference on Biomedical Wireless Technologies, Networks, and Sensing Systems, 2011.

[17]   E. Martin, "Real Time Patient's Gait Monitoring through Wireless Accelerometers with the Wavelet Transform", IEEE Conference on Biomedical Wireless Technologies, Networks, and Sensing Systems, 2011.

[18]   C. Rose, R. Yates, "Location uncertainty in mobile networks: a theoretical framework" IEEE Communications Magazine, February 1997, p. 94-101.

[19]   C. Rose, R. Yates, "Minimizing the average cost of paging under delay constraints". ACM wireless Networks, 1997, v1, n2, p. 211-219.

[20]   A. Papoulis, "Probabilty, random variables, and stochastic processes". New York, McGraw-Hill, 3rd edition, 1991.

[21]   S. Rappaport, "Blocking, hand-off and traffic performance for cellular systems with mixed platforms" IKE Proceedings I, 1993, v. 140, n 5, p. 389.

[22]   D. Kragic, A. Miller, P. Allen, "Real time tracking meets online grasp planning", Proceedings - IEEE International Conference on Robotics and Automation, 2001, v 3, p 2460-2465

[23]   C. Rose, "Minimizing the average cost of paging and registration: a timer-based method" Wireless Networks, 1996, v2, n2, p. 109-116.

[24]   E. Martin, et al., "Enhancing Context Awareness with Activity Recognition and Radio Fingerprinting", IEEE International Conference on Semantic Computing, 2011.

[25]   E. Martin, et al., "Linking Computer Vision with off-the-shelf Accelerometry through Kinetic Energy for Precise Localization", IEEE International Conference on Semantic Computing, 2011.

[26]   E. Martin, T Lin, "Probabilistic Radio Fingerprinting Leveraging SQLite in Smart Phones", IEEE International Conference on Signal Processing Systems, 2011.

[27]   E. Martin, R. Bajcsy, "Enhancements in Multimode Localization Accuracy Brought by a Smart Phone-Embedded Magnetometer", IEEE International Conference on Signal Processing Systems, 2011.

[28]   E. Martin, "Multimode Radio Fingerprinting for Localization", IEEE Conference on Wireless Sensors and Sensor Networks, 2011.

APPENDIX I. MATHEMATICAL EXPRESSION FOR THE DIMENSIONALITY THRESHOLD.

Considering a basic classification of motion processes for mobile communications networks users, based on the comparison of their call arrival rates, analyzing the behaviour of their paging costs for one, two and three dimensions of the movement, the point where the roles between the call arrival rate and the movement dimension are exchanged is located at the crossing of the costs curve of the two-dimensional one, characterized by low call arrival (λ_2), with that of the three-dimensional one, characterized by high call arrival rate (λ_3). Mathematically, the equation describing that crossing is:

$$\left(\frac{Dif_2}{\lambda_2} \cdot t \right)^{2/2} = \left(\frac{Dif_3}{\lambda_3} \cdot t \right)^{3/2} \qquad (12)$$

From (12), obtaining the time solution:

$$t = \frac{\left(\dfrac{Dif_2}{\lambda_2}\right)^2}{\left(\dfrac{Dif_3}{\lambda_3}\right)^3} = \frac{\left(\dfrac{1}{CMR_2}\right)^2}{\left(\dfrac{1}{CMR_3}\right)^3} \tag{13}$$

Therefore, naming that time interval as Dimensionality Threshold, the final expression is the following:

$$D_{TH} = \frac{\left(CMR_h\right)^3}{\left(CMR_l\right)^2} \tag{14}$$

9

BUFFER MANAGEMENT FOR PREFERENTIAL DELIVERY IN OPPORTUNISTIC DELAY TOLERANT NETWORKS

G.Fathima[1], R.S.D.Wahidabanu [2]

[1]Adhiyamaan College of Engineering, Hosur, TamilNadu, India
fathima_ace@yahoo.com
[2]Govt. College of Engineering, Salem, TamilNadu, India
drwahidabanu@gmail.com

ABSTRACT

Delay Tolerant Networks (DTN) present many challenges that are not present in traditional networks. Many stem from the need to deal with disconnections which directly impacts routing and forwarding. However as these networks enable communication between wide range of devices, there are secondary problems that routing strategies may need to take care of such as to deal with limited resources like buffer, bandwidth, power. Most of the routing protocols in DTN assume that the buffer size as infinite which is not the case in reality. In resource constrained environment, buffers will run out of capacity at certain point of time. Moreover due to mobility of the nodes and limited bandwidth, it is not possible to transmit all messages a node has during the short available period of contact. Consequently an efficient buffer management policy is required under resource constrained DTNs. Further, DTNs can be used to support several asynchronous applications simultaneously. Each application may have different priority. For example, an emergency alert in monitoring application is more important than the regular data. Such environment stimulates the need to introduce priority to messages. Therefore in this paper, a policy is proposed which performs buffer management with prioritization. The proposed approach with epidemic routing is evaluated through simulation and compared with other policies. It is shown that the proposed approach results in performance improvement to epidemic routing with preferential delivery.

KEYWORDS

Delay Tolerant Networks, Opportunistic networks, Buffer management, Prioritization of messages, Delivery ratio

1. INTRODUCTION

Delay Tolerant Networking (DTN) is a technology which supports data transfer in challenging environments where a fully connected end to end path may never exist between a source and destination. The DTN approach is well suited for deploying applications in the developing world as it allows applications to continue operating with much less infrastructure compared to more traditional networking approaches. There are many applications that make use of DTN like (i) support to low-cost internet provision in remote or developing communities [16]. (ii) in vehicular networks (VANETs) for dissemination of location dependent information (eg., local ads, traffic reports, parking information [17]). (iii) in noise monitoring , earth quake monitoring. The details of Delay Tolerant Network architecture are available in [11]. From the literature survey [9], [13], [15], [22], [24], [25] it is understood that a large amount of research has been performed in developing efficient routing algorithms for DTNs. However, it is observed that flood-based routing protocols perform poorly when resources like buffer and bandwidth are limited.

DTNs operate with the principle of store, carry and forward. In order to cope with long disconnection, messages must be buffered for long period of time. It implies that intermediate nodes require enough buffer space to store all messages that are waiting for future communication opportunities. Moreover to achieve high delivery probability, messages are replicated to each and every node they encounter. The combination of long term storage and extensive message replication performed often by many DTN routing protocols imposes a high storage overhead on wireless nodes. Moreover, bundles which are application-level data units can often be large. In this context, it is evident that buffers will run out of capacity at certain point of time.

The next important resource is the contact capacity. i.e, how much data can be exchanged between nodes. This depends on both link technology and the duration of contact. Even if duration is precisely known, it may not be possible to predict the capacity due to fluctuation in the data rate. When the number of messages to be transmitted is very small compared to the capacity of the contact in the networks then, all messages will get transmitted when nodes come in contact with each other. Here the order of transmission is not an issue. But if the number of messages to be transmitted is more than the capacity of contact then, it is not possible to transmit all messages.

Further, transmission takes place when nodes come into each other's communication range. The node has to decide which of the messages to be transmitted among the messages those are available in the buffer. Despite inherent delay tolerance of most DTN driving applications, there can be situations where some messages may be more important than the other. For e.g., in VANETs it is reasonable to assume that an accident notification message is more important than a chat message or advertisements of nearby shops. Under such requirement, a different forwarding policy will be needed to serve different types of traffic. Consequently it would be necessary to prioritize messages and ensure that they get best possible service. So given the network limitations, the key question to be answered is, how to prioritize the messages and schedule them so that messages are delivered preferentially. In this paper, an adaptive buffer management policy with prioritization is proposed which takes care of both: which messages are to be transmitted when a new contact arises and which messages are to be dropped when buffer is full. The proposed policy does selective dropping and scheduling. It considers the lifetime as well as the priority of the messages in making such decisions. The proposed approach with epidemic routing is evaluated through simulation and compared with other policies. It is shown that the proposed approach results in performance improvement to epidemic routing with preferential delivery.

This new policy is evaluated using ONE simulator and compared with other dropping and forwarding policies. ONE is an opportunistic Network Environment Simulator which is designed specifically for DTN environment. The remainder of this paper is organized as follows: Section 2 gives a background of DTN routing mechanism and the previous work related to buffer management is discussed. Section 3 discusses about the proposed method of buffer management. The simulation setup and the results are discussed in section 4. Section 5 concludes the paper.

2. BACKGROUND AND RELATED WORK

Comprehensive study of different routing mechanism is important to understand the design of DTNs. There are various routing protocols available for DTN, the details of which are available in [9], [13], [15]. They differ in the knowledge that they use in making routing decisions and the number of replication they make. The various DTN protocols are Direct Delivery, First Contact, Epidemic [4], [5], Spray and Wait [3], [21], PRoPHET [2], and MaxProp [10] routing. Among the above mentioned protocols, the first four protocols are simple routing protocols which do not require any knowledge about the network. The latter two protocols use some extra information to make decisions on forwarding. Further based on the replication, the DTN routing protocols can be classified as those that replicate multiple copies and those that forward only a single copy. The protocols like Direct Delivery and First Contact routing are single copy protocols where only one copy per message is routed. Therefore, the buffer requirement and their utilization is less in these protocols. The protocols like Epidemic, Spray & Wait and PRoPHET routing are multi-copy protocols and therefore they require more buffer space and their utilization is observed to be maximum. This is an interesting case where much research is to be carried out to yield good performance when the resources are constrained. The comparison of various DTN routing protocols and their buffer utilization has been discussed in [26].

It is necessary to understand the impact of buffer size on performance, as this resource is limited in reality. Epidemic routing is chosen as baseline for evaluation as this routing is based on flooding and requires huge buffer space. Epidemic routing floods each message throughout the network through its neighbours to achieve high delivery probability. As it relies on buffer to have a copy of every message at every node, buffer size has significant impact on delivery probability. A number of studies also have clearly shown that Epidemic routing has minimum delivery delay under no buffer and bandwidth constraints but performs poorly under constrained environments. The studies in [5] illustrate how the buffer constraints affect the performance of DTN routing severely.

Buffer management is a fundamental technology which controls the assignment of buffer resources among different traffic classes and aggregation of the same according to certain policies. An efficient buffer management policy is required to decide at each step which of the messages are to be dropped when buffer is full and which of the messages are to be transmitted when bandwidth is limited irrespective of the routing algorithms used. Table 1 shows the existing replication based DTN routing protocols and their assumption on availability of buffer and bandwidth; both being either limited or unlimited.

Table 1. Assumption of DTN Routing Protocols

DTN Routing Protocols	Buffer	Bandwidth
Epidemic, Spray & Wait	Unlimited	Unlimited
ProPHET	Limited	Unlimited
MaxProp	Limited	Limited
RAPID	Limited	Limited
Optimal Buffer Management (with Epidemic Routing)	Limited	Limited
Prioritized Epidemic Routing	Limited	Limited

The protocols like Direct Delivery and First Contact routing are single copy protocols where only one copy per message is routed in FCFS order. i.e., the messages are transmitted in the order in which they were stored in the buffer. Among the multi-copy protocols, epidemic and Spray & Wait routing also uses FCFS forwarding policy. PRoPHET routing makes forwarding decision based on delivery predictability of the destination. It needs history of past encounters for calculation of delivery predictability. MaxProp routing assigns priorities to the messages based on hop count and delivery likelihood. Estimation of delivery likelihood is done based on historical data. It forwards the messages with high priorities when a contact arises. RAPID protocol [7] derives the per-packet utility function from administrator-specified routing metric. It forwards the messages with highest utility value first. Similarly, the Optimal policy in [18],

[20], [23] derives per-message utility function from statistical learning and the message with smallest utility is dropped when the buffer is full and message with highest utility is scheduled first for transmission. In Prioritized Epidemic Routing [19], each bundle is assigned a drop priority and transmit priority which is based on hop count. i.e., the number of hops the bundle has traversed thus far. The transmission and dropping is done based on the priority. The approach presented in this paper differs from the above mentioned works in considering the traffic class and lifetime of the bundles.

The simple dropping policy used in many networks is Drop tail policy. Apart from drop tail policy there are other policies proposed in the literature [1], [6] where an arriving packet is always accepted if there is an empty buffer. Else it is accepted by dropping another packet. In general, policies which can accept an arriving packet by dropping another packet from the system are known as push – out policy. Such policies are Drop First, Drop Last, and Drop Random. Upon arrival of a packet the system can decide to either accept the packet or reject it or accept it and drop another packet based on the policy. Therefore the goal is to determine the policy which maximizes the overall throughput or equivalently minimize the overall loss probability of high priority messages.

Though a number of scheduling policies are possible, FCFS is the simple policy which is easy to implement. As long as the contact duration is long enough to transmit all messages a node has, FCFS is a very reasonable policy. However if the contact duration is limited, the policies, FCFS and drop tail are sub-optimal as it does not provide any mechanism for preferential delivering or storing of high priority messages. Considering the above said problems, the proposed policy attempts to differentiate traffic based on Class-of -Service (CoS) and provide better levels of service in a best–effort environment. Thus the proposed policy is more advantageous in emergency applications as it does preferential delivery.

3. PROPOSED SYSTEM

3.1. Motivation

Most of the existing routing protocols offer best effort service. There is one fundamental limitation of best effort method being used: it makes no attempt to differentiate between the traffic classes that are generated by different hosts. But to provide different services to different applications, it is necessary to differentiate traffic classes. . The unpredictable and bursty nature of DTN makes it necessary to manage the buffer. A final motivation for adding a service class to the DTN is to provide a means by which applications that are not intrinsically delay tolerant can still be supported by DTN deployment. More specifically, some applications that use DTN service require preferential delivery of certain messages. For e.g, field agents wish to communicate their findings, regarding environment hazards to other field agents which are more important than the regular findings. Moreover there may be some messages in the buffer whose lifetime is small and retaining them may not be useful as the time of next available contact is not known. Therefore a buffer management mechanism is required which is capable of differentiating the traffic and to transmit and drop messages so as to maximize the delivery ratio or minimize end to end latency.

3.2. Network Model

The network that is considered can be characterised as partially connected with low node density and high node mobility. The movement inherent in the nodes themselves is exploited to deliver the messages when the network is partially connected which is referred as opportunistic DTN. It is assumed that no knowledge about the network is known a priori and no infrastructure exists to provide connectivity. Assume that N is the total number of nodes in the network. Each of these nodes has a buffer, which can store either messages belonging to other nodes or messages generated by itself. Each message is destined to one of the nodes in the network and has a Time-To-Live (TTL) value. Once the TTL value expires, the message is no more useful to the application and it is dropped from the buffer. In the context of DTN, message transmission occurs only when nodes encounter each other. Consider a node which acts as an intermittent to several flows. That is the messages from several senders enter the node at various instances. Since node mobility is assumed, the node may accept messages either from other routing nodes or directly from senders. In the first case, the messages arrive back to back with constant inter arrival times and in the second case in a stochastic manner. The node then has to keep them until a connection opportunity occurs or until its storage space is full.

3.3. Queue Model

In DTN, bundle protocol is used for transfer of messages. A bundle is a protocol data unit of the DTN bundle protocol [8]. Bundle Processing Control Flags Bit is used to differentiate the traffic through Class-of-Service (CoS) field. The Lifetime field available in the primary bundle block gives the expiration time. This information is used for prioritization.

It is assumed that there are three priority classes of traffic: bulk, normal and expedited. Bulk messages have lowest priority, normal messages have medium priority and expedited messages have high priority. The policy gives preference to expedited messages. In the proposed approach, the available buffer is divided into many queues to hold the incoming bundles. Separate queue is maintained for each class of service as shown in Figure 1. At this point the size of the queues is not determined. Their sizes can be either defined at the beginning or varied as the correlation of traffic changes. The goal is to determine how the buffers are best shared among messages of different classes, so that the overall delivery ratio is maximized.

Assume that each node has a buffer B of size $b = n(B)$ which is logically divided into three queues: B_1, B_2, B_3 to accommodate high, medium and low priority bundles respectively such that $B = \{B_1 \cup B_2 \cup B_3\}$. The size of B_1, B_2, B_3 is b_1, b_2, b_3 respectively such that $b = b_1 + b_2 + b_3$. To avoid the complete negligence of medium and low priority traffic, a minimum size q_{min} is reserved for medium and low priority queues. The value of q_{min} is set dynamically according to the requirements of the application.

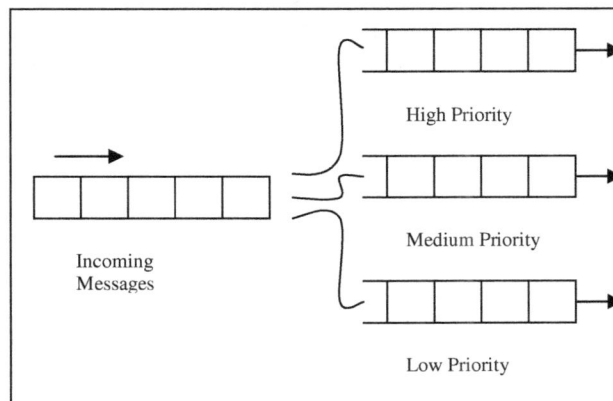

Figure 1. Maintaining Priority Queue

3.4. Buffer Management System

The proposed system comprises of (i) *Bundle classifier* which classifies the bundles according to their priority as soon as they arrive and stores them in appropriate queue, (ii) *Bundle scheduler* which is invoked when the contact opportunity arises and schedules the bundles based on the policy (iii) *Bundle dropper* which is invoked when buffer is full and drops the bundle according to the policy. This system is illustrated in Figure 2.

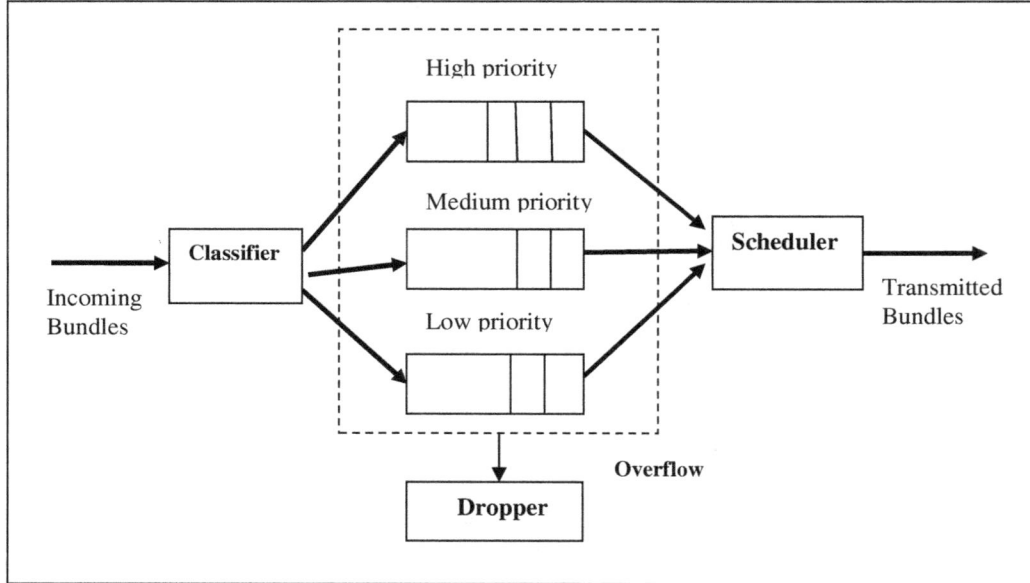

Figure 2. System Overview

Bundle classifier

Each bundle in the buffer has a set of information stored with it such as source id, traffic class and Time-To-Live(TTL). Initially bundles are classified based on their Class of Service and stored in appropriate queue as and when they arrive, by the Bundle Classifier. The service class can be specified by the application. The Bundle Classifier is a function of newly arrived bundle b_{new} such as

$$f(b_{new}) = \begin{cases} B_1 \cup \{b_{new}\} & \textit{if } b_{new} \textit{ is of high priority} \\ B_2 \cup \{b_{new}\} & \textit{if } b_{new} \textit{ is of medium priority} \\ B_3 \cup \{b_{new}\} & \textit{if } b_{new} \textit{ is of low priority} \end{cases}$$

where B_1 contains $\{b_{11}, b_{12}, ..., b_{1b1}\}$ which are high priority messages, B_2 contains $\{b_{21}, b_{22}, ..., b_{2b2}\}$ which are medium priority messages, B_3 contains $\{b_{31}, b_{32}, ..., b_{3b3}\}$ which are low priority messages.

Bundle dropper

When the entire buffer is full, some of the bundles should be dropped to give room for new bundles. So once the buffer is full, the Bundle dropper is invoked. The Bundle dropper drops the low and medium priority bundles to give room for high priority bundles. A bundle is dropped automatically when the TTL expires. It is also taken care that a node should not drop its own bundle (source) to give room for newly arrived bundles. The idea of giving priority to source bundles has been proposed in [19], and was shown to improve the average delivery ratio. So the same idea is followed here. Bundle dropping is a function which identifies the bundle to be dropped according to the proposed policy. The bundle drop procedure is as follows:

Case 1:

When high priority bundle arrives, the bundle to be dropped b_{drop} is identified as follows:

$$\begin{cases} if \ b_3 > q_{\min} \ , b_{drop} \in B_3 \ with \ least \ TTL \ value \\ else \ if \ b_2 > q_{\min} \ , b_{drop} \in B_2 \ with \ least \ TTL \ value \\ else \ \ b_{drop} \in B_1 \ with \ least \ TTL \ value \end{cases}$$

Case 2:

When medium priority bundle arrives, the bundle to be dropped b_{drop} is identified as follows:

$$\begin{cases} if \ b_3 > q_{\min} \ , b_{drop} \in B_3 \ with \ least \ TTL \ value \\ else \ b_{drop} \in B_2 \ with \ least \ TTL \ value \end{cases}$$

Case 3:

When Low priority bundle arrives, the bundle to be dropped b_{drop} is identified as follows:

$$b_{drop} \in B_3 \ with \ least \ TTL \ value$$

Bundle scheduler

When two nodes come into the communication range of each other, they start exchanging messages. Short duration of contact between the nodes and finite bandwidth may not allow the node to transmit all the messages that are available in the buffer. In such cases the order in which the messages are transmitted is significant. Bundles are transmitted according to their priority and ordered based on expiration time. Bundle scheduler transmits the bundles from high priority to low priority in a round robin fashion. The bundle to be transmitted b_{sch} is identified as follows:

$$until \ nodes \ are \ in \ communication \ range$$

$$\begin{cases} while \ high \ priority \ queue \ is \ not \ empty, b_{sch} \in B_1 \\ while \ medium \ priority \ queue \ is \ not \ empty, b_{sch} \in B_2 \\ while \ low \ priority \ queue \ is \ not \ empty, b_{sch} \in B_3 \end{cases}$$

It should be noted that irrespective of the scheduling policy adopted, the messages whose destination encountered are the first to be transmitted and the same may be deleted from the buffer. Nodes do not delete messages that are forwarded to other nodes (i.e., not to destination) as long as there is sufficient space available in the buffer.

Performance of DTN is measured in terms of average delivery ratio and average delivery delay. The average delivery ratio is defined as ratio of number of messages delivered to the destination and the total number of messages sent by the sender. The average delivery delay is measured as the average of the time taken to reach from source to destination by all messages. Both the metric equations are shown below:

$$\text{Delivery ratio} = \frac{\text{Number of messages delivered}}{\text{Total number of messages sent by the sender}} \qquad \ldots \quad (1)$$

$$\text{Delivery delay} = \text{Average (the time taken to reach from source to destination by all messages)} \qquad \ldots \quad (2)$$

4. SIMULATION RESULTS AND ANALYSIS

4.1. Simulation Environment

To evaluate the proposed scheme, the ONE Simulator [14] has been used. ONE is an Opportunistic Network Environment simulator which is designed specifically for DTN environment. It is a discrete event based simulator. It is a Java-based tool which provides DTN protocol simulation capabilities in a single framework. A detailed description of this simulator is available in [12]. The Mobility model used is Random Way Point (RWP) model. It is the model in which nodes move independently to a randomly chosen destination. As the network with random behaviour is considered, Epidemic routing is used as the routing algorithm.

The simulation environment consists of sparsely distributed mobile nodes and they communicate when they are in the communication range of one another. The settings of the group of nodes like buffer size transmit range, transmit speed, group speed, wait time, number of nodes in the group are set as mentioned in the Table 2.

Table 2. Parameters

Parameters	Values
Number of Nodes	100
Transmit Range(m)	250
Transmit speed (Mbps)	2
Node Speed (km/hr)	10-60
Message size (MB)	1-2
TTL of message (min)	30
Buffer size (MB)	15
Simulation Time (s)	43000

4.2. System Evaluation and Analysis

The performance of epidemic routing under different buffer management policies is compared in terms of metrics like delivery probability and average delivery latency. Simulation results for different dropping policies with respect to delivery probability and delivery delay are shown in the Figure 3 and Figure 4 respectively. The different dropping policies that are compared are Drop Old (DO), Drop Young (DY), Drop Random (DR) and the Prioritized Policy (PP). It can be observed from the result in Figure 3 that as and when the traffic load increases, the delivery probability decreases. At the same time it also shows that there is not much difference in delivery ratio upon incorporating the new prioritized policy compared with other policies. But the prioritized policy guarantees the delivery of high priority messages first. Similarly it can be inferred from the result in Figure 4 that the delivery latency increases rapidly irrespective of the policies. But the proposed policy guarantees the delivery of high priority messages with least delay.

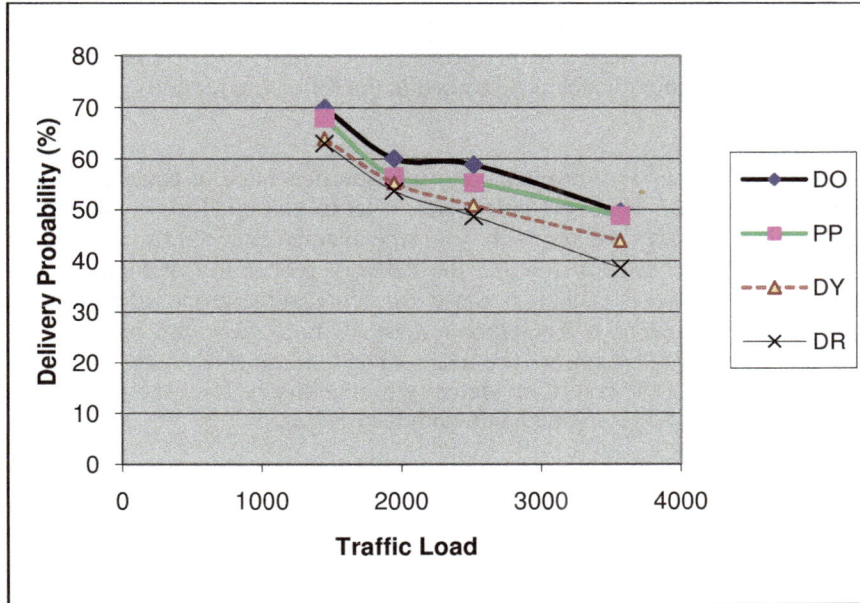

Figure 3. Delivery Probability as a function of Load

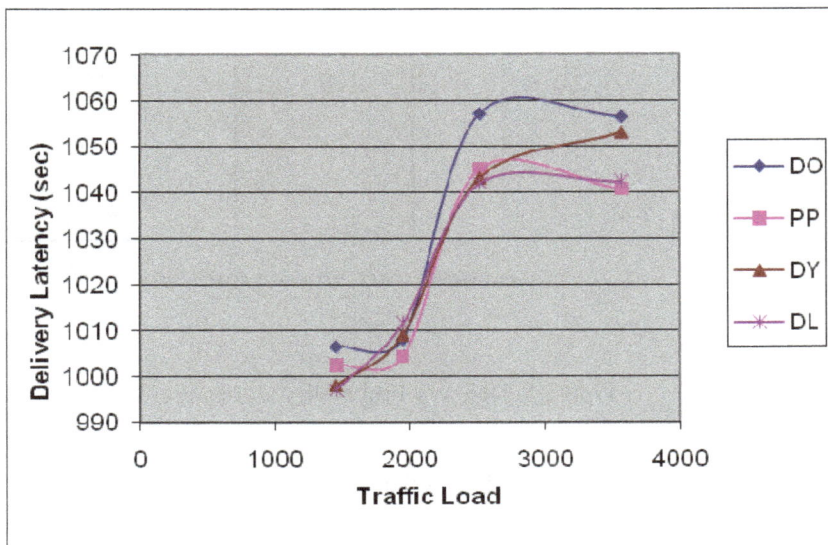

Figure 4. Delivery delay as a function of Load

It has been shown in [1] that DO policy gives better performance in terms of delivery ratio among the different drop policies in epidemic routing. The simulation results shown in Figure 3 support it. The rationale behind this result is that an old message is likely to be duplicated to more nodes and discarding a copy of it may not decrease the overall delivery ratio much. The new prioritized policy which combines the lifetime and the priority of the messages do not decrease the delivery ratio compared to DO policy but guarantees delivery of high priority messages first. The rationale behind this result is that messages with less remaining lifetime may get automatically removed from the buffer when the lifetime expires. So forcing such messages to drop will not decrease the delivery ratio. Moreover messages with high priority are forwarded first. So they have more chances of earliest delivery than other messages. The system is further evaluated to check the performance behaviour of different priority messages at different rate of generation which is discussed in the following section.

Scenario 1:

In scenario 1, messages with different priorities are generated at equal rate and their delivery probability is observed. Here, the buffer size is set to unlimited to accommodate all messages and bandwidth is limited. The result shown is the average of several simulation runs. The result in the graph of Figure 5 confirms that the delivery probability of high priority messages is higher than other messages. This is because according to the proposed policy, all high priority messages are scheduled first. Therefore almost all messages with high priority reach their destination before the TTL expires. The result when buffer size is limited is not shown here due to space limitations. In the result, the delivery probability of low priority messages is affected due to short duration of contact and bandwidth restrictions.

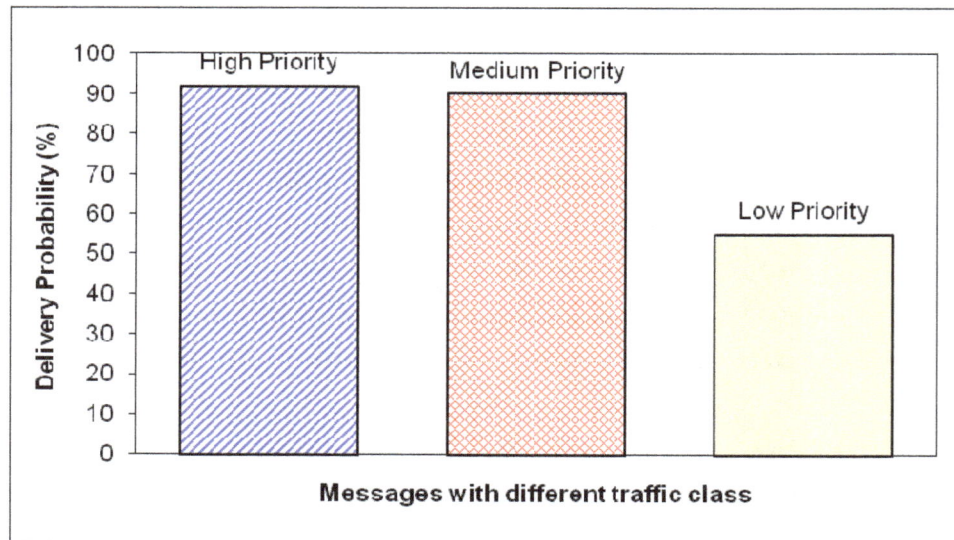

Figure 5. Delivery Probability at scenario-1

Scenario 2:

In scenario 2, messages are generated such that rate of high priority messages is more than the rate of low & medium priority messages (i.e., high priority messages are doubled that of low and medium priority messages). In scenario 2 and 3, the bandwidth and the buffer size are set to be limited. When the load of high priority messages is increased, the delivery probability of low priority messages gets decreased. It can be observed in the result of Figure 6. The rationale behind this result is that only low and medium priority messages are dropped to give room for high priority messages when there is an overflow. Moreover due to bandwidth limitations, almost all high priority messages are transmitted and only few low priority messages are transmitted. Therefore their delivery probability decreases.

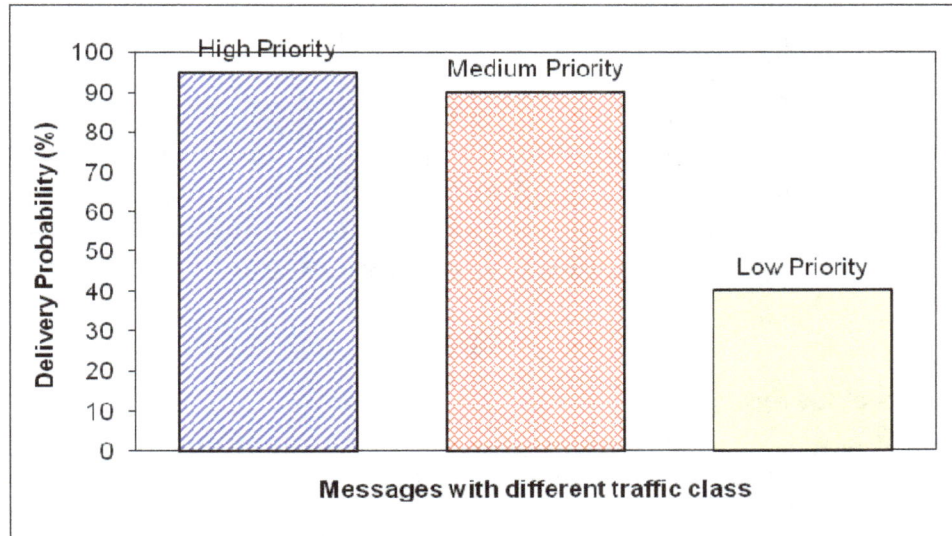

Figure 6. Delivery Probability at scenario-2

Scenario 3 :

In scenario 3, messages are generated such that rate of high priority messages is less than the rate of low & medium priority messages (i.e, high priority messages are halved that of low & medium priority messages). Even when there is increase in traffic load of medium and low priority messages, the delivery probability of high priority messages is not affected and remains higher. The rationale behind this result is that, as high priority messages are less, the overflow is due to only low and medium priority messages. In this scenario according to the policy, only low priority messages are dropped to give room for medium priority messages. Therefore the delivery probability of high and medium priority messages is not affected and the delivery probability of low priority messages is much lesser. It is inferred from the result shown in Figure 7.

It is observed that if a bit of low priority message loss is compromised, then high priority message loss can be reduced up to a greater extent. This improvement of the performance of high priority message loss at a little compromise of low priority message loss gives a special merit to the proposed policy. This ensures that the high prioritized traffic is forwarded with least delay and least likelihood of being dropped due to buffer overflow. It has the limitation that when the proportion of high priority messages is increased rapidly, high priority messages are less likely to get a low priority message to push out and that causes high priority loss to increase. Practically number of high priority messages will be a limited one. However using proposed policy one can alliance order of magnitude improvement in the high priority performance at the cost of moderate medium and low priority performance.

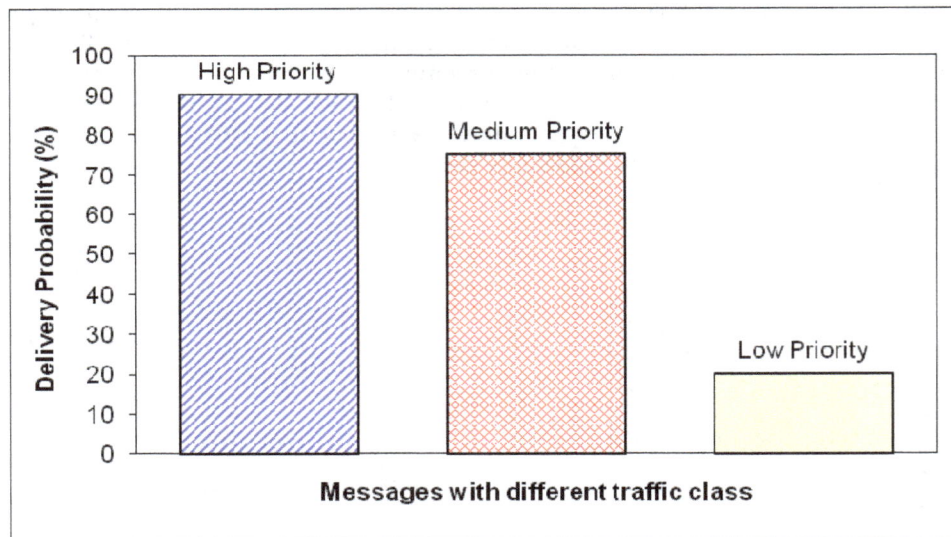

Figure 7. Delivery Probability at scenario-3.

4.3. Merits of the System

The work targets on application that requires preferential delivery in opportunistic DTN environment. By providing differentiated service based on the class, the best effort service has been enhanced. When compared to other approaches of [18], [19], [20], [23], the proposed approach has the credit of state less approach that minimizes the need for nodes in the network to remember anything about flows. It makes the proposed approach more practical to implement and more scalable. The messages are marked in a way that describes the service level that they should receive. Moreover it has less overhead than other approaches as there is no exchange of control traffic before bundle exchange. Since the approach is successful for opportunistic environment which is non-deterministic in nature, it can be made easily applicable for deterministic DTN. The proposed approach is more advantageous when there is strict constrains on resources like buffer and bandwidth.

5. CONCLUSION

Most of the DTN routing protocols operate with the assumption of infinite buffer and bandwidth. However, these resources are limited in a realistic environment. Moreover in this paper, the DTN environment considered is highly mobile and opportunistic in nature which limits the duration of contact. Therefore the work focused on effective buffer management. The approach presented in this paper prioritizes the traffic based on Class-of-Service and expiration time and performs the scheduling and dropping based on priority. It satisfies the application-specific requirements as the service required can be specified by the application. The proposed approach is validated through simulation. The results illustrate that the approach presented performs more or less equally to other policies in terms of delivery ratio with preferential delivery of high priority messages. The proposed policy is more suitable and advantageous in strict resource constrained environment with emergency applications. So it can be used in vehicular networks where accident notification is more important than other messages. The Fair queuing with dynamically assigned weights, can be utilized for controlling the quality of service. Thereby it addresses the integration of QoS in the DTN framework providing a bound on performance metrics like delay or throughput. Apart from delivery ratio, the other metrics like loss probability and power consumption can be considered for optimization. Finally, the work can be extended by considering energy consumed during the transmission of the messages which is carried as future work.

REFERENCES

[1]   A. Davids, A. H. Fagg, and B. N. Levine, (2001) "Wearable Computers as Packet Transport Mechanisms in Highly- Partitioned Ad- Hoc Networks," Proc. Int'l. Symp Wearable Comp., Zurich.pp. 141-148.

[2]   A. Lindgren, Oluv Schelen, (2004) "Probabilistic Routing in Intermittently Connected Networks," Springer LNCS, Vol 3126. pp 239-254.

[3]   A. Spyropoulos et al., (2005) "Spray and Wait: An Efficient Routing Scheme For Intermittently Connected Mobile Networks," ACM SIGCOMM Wksp. Delay Tolerant Networking (WDTN-05).

[4]   Alan Demers, Dan Greene, Carl Hauser, Wes Irish, John Larson, Scott Shenker, Howard. Sturgis, Dan Swinehart, and Doug Terry. (1987) "Epidemic algorithms for replicated database maintenance." In Proceedings of the 6th ACM Symposium on Principles of Distributed Computing, pp 1-12.

[5]   Amin Vahdat and David Becker. (2000) "Epidemic Routing for partially connected ad hoc networks", Technical report CS- 200006, Duke University. http://issg.cs.duke.edu/epidemic/

[6]   Anders Lindgren, Kaustubh S. Phanse, (2006) "Evaluation of Queuing Policies and Forwarding Strategies for Routing in Intermittently Connected Networks" in proceedings of IEEE COMSWARE '06, pp 1-10.

[7]   Aruna Balasubramanian, Brian Neil Levine and Arun Venkataramani, (2007) "DTN Routing as a Resource Allocation Problem", ACM SIGCOMM, Computer Communication Review, Vol. 37, No.4.

[8]   Bundle Protocol specification , (2007) IETF, RFC 5050.

[9]   Evan P.C. Jones, Paul A.S.Ward, (2007) " Practical Routing in Delay Tolerant Networks" , IEEE Transactions on Mobile Computing, Vol. 6, No. 8,pp. 943-959.

[10]   J. Burgess, B. Gallagher, D.Jensen and B.N. Levine, (2007) "MaxProp: Routing for vehicle-Based Disruption-Tolerant Networks", in Proceedings of IEEE Infocom, pp 1-11.

[11]   Kevin Fall, (2007) "A delay-Tolerant Network Architecture for challenged Internet ", IETF, RFC 4838.

[12]   Keranen, A. (2008) "Opportunistic Network Environment Simulator," Special Assignment report, Helsinki University of Technology, Department of Communication and Networking.

[13]   Sushant Jain, Kevin Fall, Rabin Patra, (2004) "Routing in Delay Tolerant Network," in proceedings of ACM- SIGCOMM '04.

[14]   TKK/COMNET. Project page of the ONE simulator. 2008 http://www/netlab.ttk.fi/tutkimus/dtn/theone,

[15]   Zhensheng Zhang, San Diego Research Center , (2006) "Routing In Intermittently Connected Mobile AdHoc Networks And Delay Tolerant Networks: Overview and Challenges," IEEE Communications Surveys Tutorials, Vol. 8, No. 1. pp 24-37.

[16]   A. Pentland, R.Fletcher and A.Hasson, (2004) "Daknet: Rethinking connectivity in developing nations," IEEE Computer,Vol. 37,No.1 pp.78-83.

[17]   P. Basu and T.Little, (2002) "Networked parking spaces: architecture and applications," in IEEE vehicular Technology conference (VTC).

[18]   Amir Krifa, Chadi Barakat, Thrasyvoulous Spyropolous, (2008) " An Optimal Buffer Management Policies for Delay Tolerant Networks", in Proceedings of IEEE SECON 2008, pp. 260-268.

[19]   Ram Ramanathan, Richard Hansen, Prithwish Basu, Regina Rosales-Hain, Rajesh Krishnan, (2007) "Prioritized Epidemic Routing for Opportunistic Networks," MobiOpp '07.

[20]   David Hay, Paolo Giaccone, (2009) "Optimal Routing and Scheduling for Deterministic Delay Tolerant Networks", in Proceedings of the Sixth international conference on Wireless On-Demand Network Systems and Services- 2009

[21]   Thrasyvoulous Spyropolous, Cauligi S. Raghavendra, (2008) "Efficient Routing in Intermittently Connected Mobile Networks: the Multi-Copy Case," in IEEE/ACM Trans. On Networking, Vol 16. No.1 pp. 63-76.

[22]   E.M.Daly, M.Haahr, (2010) "The Challenges of Disconnected Delay-Tolerant MANETs", Elsevier Ad Hoc Networks Journal, vol. 8, no. 2, pp. 241–250.

[23]   Amir Krifa, Chadi Barakat, Thrasyvoulous Spyropolous, (2008) "An Optimal Joint Scheduling and Drop Policy for Delay Tolerant Networks", IEEE WoWMoM -08, pp. 1-6.

[24]   "Delay tolerant networking research group." [Online]. Available: http://www.dtnrg.org

[25]   S. Farrell, V. Cahill, D. Geraghty, I. Humpreys and P. McDonald, (2006) " When TCP Breaks: Delay and Disruption- Tolerant Networking", IEEE Internet Computing, Vol.10, No.4, pp 72-78.

[26]   G.Fathima, Dr. R.S.D Wahidabanu, (2010). "Comparison of DTN routing protocols and their Buffer utilization using ONE simulator," in Proceedings of International Conference on Wireless Networks- WorldComp '10, Vol.1 pp 252-256.

10

Performance Analysis of LS and LMMSE Channel Estimation Techniques for LTE Downlink Systems

Abdelhakim Khlifi[1] and Ridha Bouallegue[2]

[1]National Engineering School of Tunis, Tunisia
abdelhakim.khlifi@gmail.com
[2]Sup'Com, Tunisia
ridha.bouallegue@gmail.com

ABSTRACT

The main purpose of this paper is to study the performance of two linear channel estimators for LTE Downlink systems, the Least Square Error (LSE) and the Linear Minimum Mean Square Error (LMMSE). As LTE is a MIMO-OFDM based system, a cyclic prefix is inserted at the beginning of each transmitted OFDM symbol in order to completely suppress both inter-carrier interference (ICI) and inter-symbol interference (ISI). Usually, the cyclic prefix is equal to or longer than the channel length but in some cases and because of some unforeseen channel behaviour, the cyclic prefix can be shorter. Therefore, we propose to study the performance of the two linear estimators under the effect of the channel length. Computer simulations show that, in the case where the cyclic prefix is equal to or longer than the channel length,LMMSE performs better than LSE but at the cost of computational complexity.In the other case, LMMSE continue to improve its performance only for low SNR values but it degrades for high SNR values in which LS shows better performance for LTE Downlink systems. MATLAB Monte − Carlo simulations are used to evaluate the performance of the studied estimators in terms of Mean Square Error (MSE) and Bit Error Rate (BER) for2x2 LTE Downlink systems.

KEYWORDS

LTE, MIMO, OFDM, cyclic prefix, channel length, LS, LMMSE

1. INTRODUCTION

In order to satisfy the exponentialgrowing demand of wireless multimedia services, a high speed data access is required. Therefore, various techniques have been proposed in recent years to achieve high system capacities. Among them,we interest to the multiple-input multiple-output (MIMO).The MIMO concept has attracted lot of attention in wireless communications due to its potential to increase the system capacity without extra bandwidth [1].Multipath propagation usually causes selective frequency channels. To combat the effect of frequency-selective fading, MIMO is associated with orthogonal frequency-division multiplexing (OFDM) technique. OFDM is a modulation technique which transforms frequency selective channel into a set of parallel flat fading channels.A cyclic prefix CP is added at the beginning of each OFDM symbol to eliminate ICI and ISI. Theinserted cyclic prefix is equal to or longer than to the channel[2].

The 3GPP Long Term Evolution (LTE) is defining the next generation radio access network. LTE Downlink systems adopt Orthogonal Frequency Division Multiple Access (OFDMA) and MIMO to provide up to 100 Mbps (assuming a 2x2 MIMO system with 20MHz bandwidth).

The performance of a MIMO-OFDM communication systemsignificantlydependsupon the channel estimation. Channel estimation techniques for MIMO-OFDM systems were carried out in many articles e.g. [3] [4]. However, in most of these research works, the CP length is assumed to be equal or longer than the maximum propagation delay of the channel. But in some cases and because of some unforeseen channel behaviour, the cyclic prefix can be shorter than channel length. In this case, both ICI and ISI will be introduced and this makes the task of channel estimation more difficult. Equalization techniques that could flexibly detect the signals in both cases in MIMO-OFDM systems are discussed in [8] [9].

In this paper, we will focus on the study of the performance of LS and LMMSE channel estimation techniques for LTE Downlink systems under the effect of the channel length. The performance evaluation of the two estimators for LTE systems was discussed in many articles e.g. [10] [11]

In the rest of the paper, we give an overview of LTE Downlink system in section II. A LTE MIMO-OFDM system model is given in section III. We discuss the two linear channel estimation techniques, LS and LMMSE in section IV with their simulation results for their performances under the effect channel length given in section V. Conclusion is given in the last section.

2. OVERVIEW OF LTE DOWNLINK SYSTEM

According to [5], the duration of one frame in LTE Downlink system is 10 ms.Each LTE radio frame is divided into 10 sub-frames of 1 ms. As described in Figure 1, each sub-frame is divided into two time slots, each with duration of 0.5 ms. Each time slot consists of either 7 or 6 OFDM symbols depending on the length of the CP (normal or extended). In LTE Downlink physical layer, 12 consecutive subcarriers are grouped into one Physical Resource Block (PRB). A PRB has the duration of 1 time slot. Within one PRB, there are 84resource elements(12 subcarriers × 7OFDM symbols) for normal CP or 72 resource elements (12subcarriers × 6OFDM symbols) for extended CP.

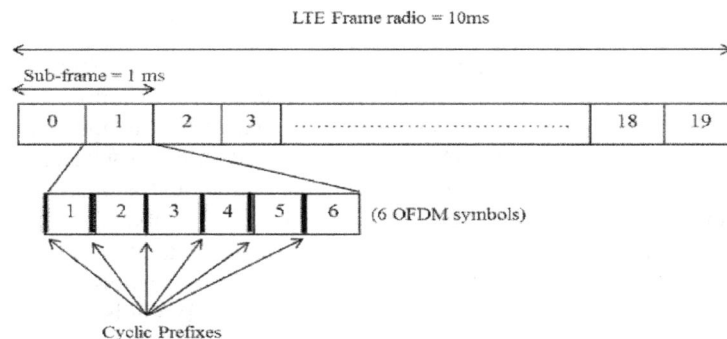

Figure 1: LTE radio Frame structure

LTE provides scalable bandwidth from 1.4 MHz to 20 MHz and supports both frequency division duplexing (FDD) and time-division duplexing (TDD). Table 1 shows the different transmission parameters for LTE Downlink systems.

Table 1: LTE Downlink parameters

Transmission Bandwitdh (MHz)	1.25	2.5	5	10	15	20
Sub-frame duration (ms)	0.5					
Sub-carrier spacing (kHz)	15					
Sampling Frequency (MHz)	1.92	3.84	7.68	15.36	23.04	30.72
FFT size	128	256	512	1024	1536	2048
Number of occupied sub-carriers	76	151	301	601	901	1201

3.DOWNLINKLTE SYSTEM MODEL

The system model is given in Figure.2. A MIMO-OFDM system with N_{Tx} transmit and N_{Rx} receive antennas is assumed.

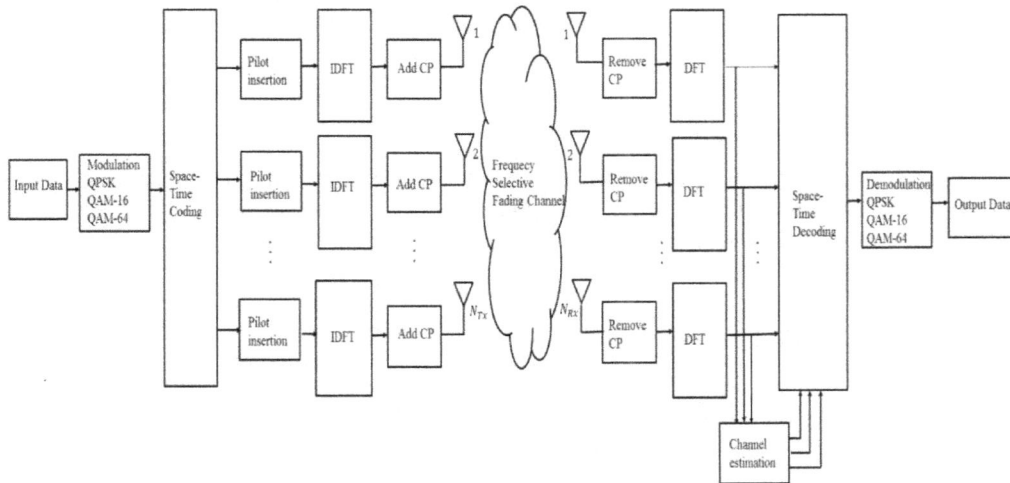

Figure 2: MIMO-OFDM system

OFDMA is employed as the multiplexing scheme in the LTE Downlink systems. OFDMA is a multiple users radio access technique based on OFDM technique. OFDM consists in dividing the transmission bandwidth into several orthogonal sub-carriers. The entire set of subcarriers is shared between different users.

Figure 3 illustrates a baseband OFDM system model.The$N$complex constellation symbols c_i are modulated on the orthogonal sub-carriers by mean of the Inverse Discrete Fourier.The$N$ subcarriers are spaced by$\Delta f = 15\ KHz$. To combat the effect of frequency-selective fading, a cyclic prefix (CP) with the length of $L_{CP}$is inserted at the beginning of each OFDM symbol.

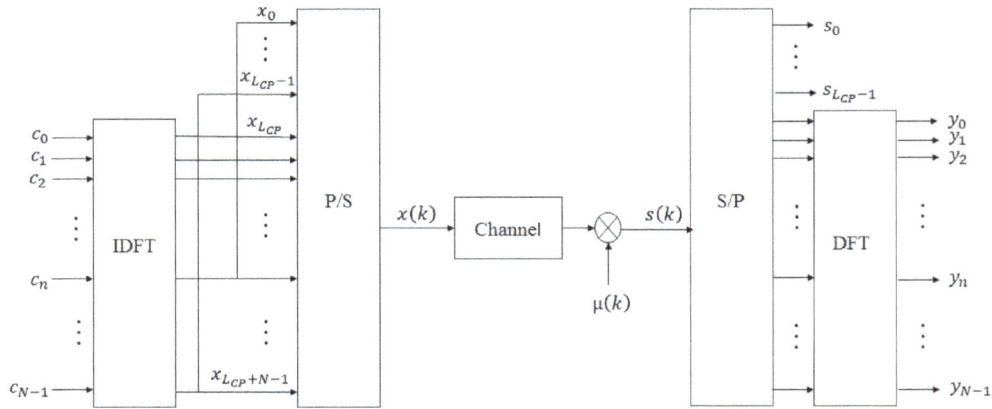

Figure 3: Baseband OFDM system

Each OFDM symbol is transmitted overfrequency-selective fading MIMO channels assumed independents of each other. Each channel ismodeled as a Finite Impulse Response (FIR) filter with Ltaps.

Therefore, we consider in our systemmodel only a single transmit and a single receive antenna. Afterremoving the CP and performing the DFT, the received OFDM symbol at one receive antenna can bewritten as:

$$Y = XH + \mu \tag{1}$$

Y represents the received signal vector; X is a matrix which contains the transmitted elements on its diagonal. H is a channel frequency response, and μ is the noise vector whose entries have the i.i.d. complex Gaussian distribution with zero mean and varianceσ_μ^2.We assume that the noise μ is uncorrelated with the channel H.

4.CHANNEL ESTIMATION

In order to estimate the channel, LTE systems use pilot signals called reference signals.When short CP is used, they are being transmitted during the first and fifth OFDM symbols of every slot. When long CP is used, they are transmitted during the first and the fourth OFDM symbols.

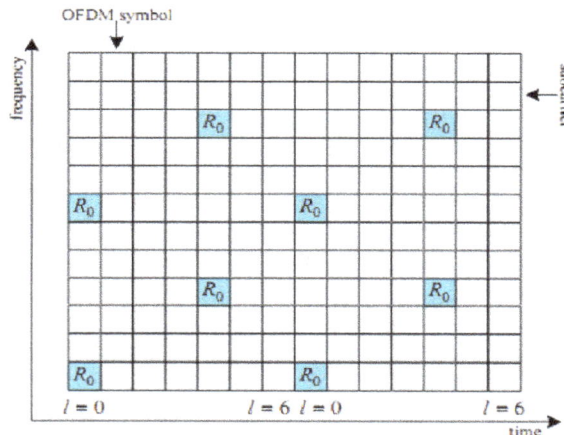

Figure 4: Downlink reference signal structure on one antenna port [5]

From (1), the received pilot signals can be written as:

$$Y_P = X_P H_P + \mu_P \tag{2}$$

$(.)_P$ denotes positions where reference signals are transmitted.
In this paper, we study the performance ofLS and LMMSE channel estimation techniques.

4.1 Least Square LS

The goal of the channel least square estimator is to minimize the square distance between the received signal and the original signal.

The least square estimates (LS) of the channel at the pilot subcarriers given in (2) can be obtained by the following equation [6]:

$$\hat{H}_P^{LS} = (X_P)^{-1} Y_P \tag{3}$$

$\hat{H}_P^{LS}$ represents the least-squares (LS) estimate obtained over the pilot subcarriers.

4.2 Linear Mean Minimum Square Error LMMSE

The LMMSE channel estimator is designed to minimize the estimation MSE.The LMMSE estimate of the channel responses given in (2) is [7]:

$$H_P^{LMMSE} = R_{HH_P}\left(R_{H_P H_P} + \sigma_\mu^2 (XX^H)^{-1}\right)\hat{H}_P^{LS} \tag{4}$$

R_{HH_P} represents the crosscorrelation matrix between all subcarriers and the subcarriers with reference signals.$R_{H_P H_P}$ represents the autocorrelation matrix of the subcarriers with reference signals.The high complexity of LMMSE estimator (4) is due to the inversion matrix lemma.Every time data changes, inversion is needed.The complexity of this estimator can be reduced by averaging the transmitted data. Therefore, we replace the term $(XX^H)^{-1}$ in (4) with its expectation$E[(XX^H)^{-1}]$.
The simplified LMMSE estimator becomes [7]:

$$\hat{H}_P^{LMMSE} = R_{HH_P}\left(R_{H_P H_P} + \frac{\beta}{SNR} I_P\right)^{-1} \hat{H}_P^{LS} \tag{5}$$

whereβ is scaling factor depending on the signal constellation (i.e$\beta = 1$ for QPSK and $\beta = 17/9$ for 16-QAM). SNR is the average signal-to-noise ratio, and I_P is the identity matrix.

5. SIMULATION RESULTS

In this section, we compare the performance of the LS and the LMMSE estimation techniques for 2×2 LTE-5MHz Downlink system under the effect of the channel length. The transmitted signals are quadrature phase-shift keying (QPSK) modulated. The number of subcarriers in each OFDM symbol is $N = 300$, and the length of CP isL_{CP}=16. 100 LTE radio frames are sent through a frequency-selective channel. The frequency-selective fading channel responses are randomly generated with a Rayleigh probability distribution.

5.1 Case with $L \leq L_{CP}$

In this case, the cyclic prefix is longer than the channel which means that ISI and ICI are completely supressed. Figure 5 and Figure 6 shows that LMMSE estimation technique is better than the LS estimator.

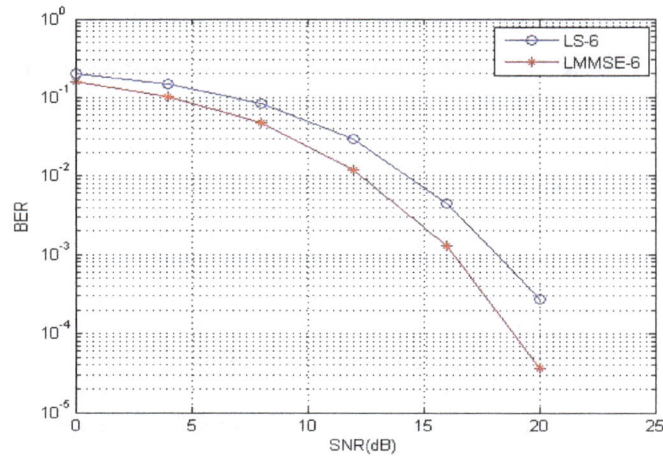

Figure 5: BER versus SNR for $L = 6$

Although, LMMSEgives the best performance but its complexity is higher due to the channel correlation and the matrix inversion lemma.

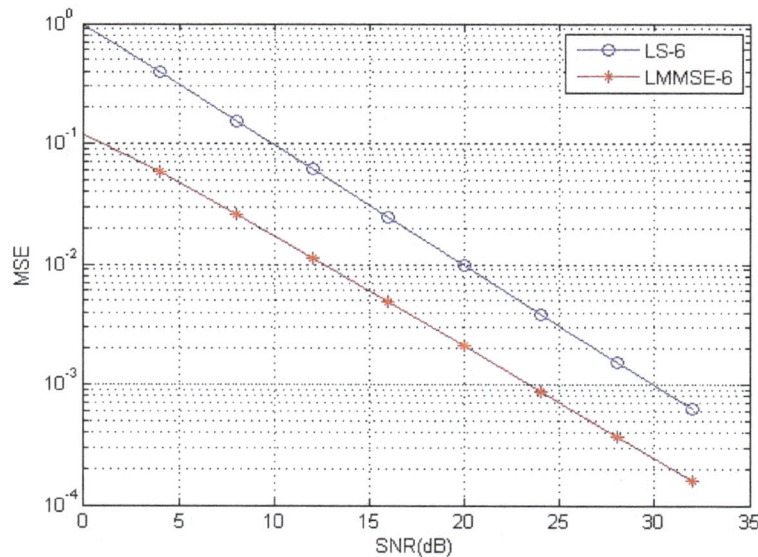

Figure 6: MSE versus SNR for $L = 6$

5.2 Case with $L > L_{CP}$

In this case, the cyclic prefix is shorter than the channel. ISI and ICI will be introduced.Figure 7 shows that more the channel is longer than CP, more the performance is lost in terms of BER at the cost of more complexity for LMMSE estimation technique.

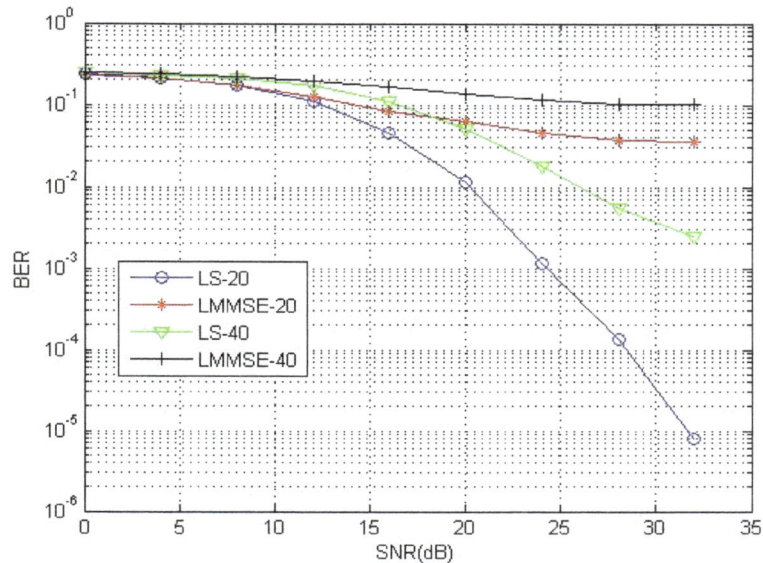

Figure 7: BER versus SNR for $L = 20$ and $L = 40$

Figure 8 demonstrates that LMMSE that;even the cyclic prefix is shorter than the channel length; LMMSE shows also better performances than LS for LTE Downlink systems but only for low SNR values. Forhigh SNR values, LMMSE loses its performance in terms of MSE and LS estimator seems to perform better than LMMSE for this range of SNR values.

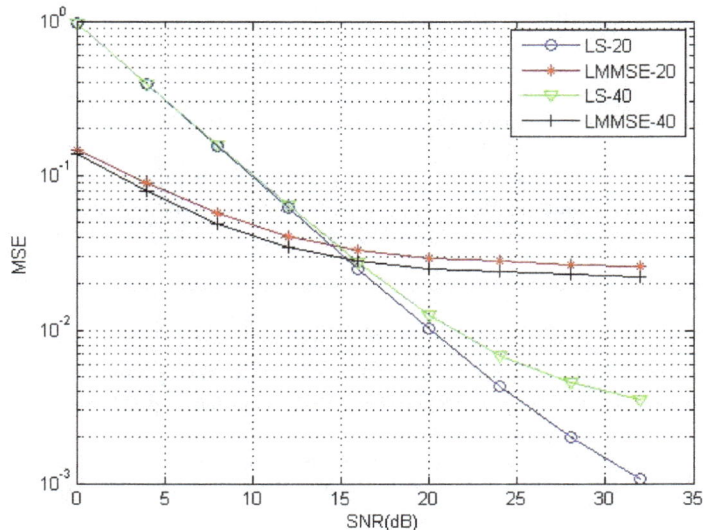

Figure 8: MSE versus SNR for $L = 20$ and $L = 40$

6. Conclusion

In this paper, we propose to evaluate the performance of LS and LMMSE estimation techniques for LTE Downlink systems under the effect of the channel length. The cyclic prefix inserted at the beginning of each OFDM symbol is usually equal to or longer than the channel length in order to suppress ICI and ISI. However, the CP length can be shorter than the channel length because of some unforeseen behaviour of the channel. Simulation results show that in the case where the CP length is equal to or longer than the channel length, the LMMSE performs better thanLS estimator but at the cost of the complexity because it depends on the channel and noise statistics. In the other case,LMMSEprovidesbetter performanceonly for low SNR values and begins to lose its performance for higher SNR values. In other hand, LS shows better performance than LMMSE in this range of SNR values.

REFERENCES

[1]   A. J. Paulraj, D. A. Gore, R. U. Nabar, and H. Bolcskei, "An overview ofMIMO communications—A key to gigabit wireless," Proc. IEEE, vol. 92,no. 2, pp. 198–218, Feb. 2004.

[2]   B. Muquet, Z. Wang, G. B. Giannakis, M. de Courville, andP. Duhamel, "Cyclic prefixing or zero padding for wireless multicarrier transmissions?" IEEE Trans. Commun., vol. 50, no. 12, pp. 2136–2148, Dec.2002.

[3]   Y. Li, N. Seshadri, and S. Ariyavisitakul, "Channel estimation for OFDM systems with transmitter diversity in mobile wireless channels," IEEE J. Sel. Areas Commun., vol.17, no.3, pp.461–471, March 1999.

[4]   D. Wan, B. Han, J. Zhao, X. Gao, and X. You, "Channel estimation algorithms for broadband MIMO-OFDM sparse channel," Proc.14th IEEE Int. Symp. on Personal, Indoor and Mobile Radio Communications, pp.1929–1933, Beijing, China, Sept. 2003.

[5]   3GPP, "Evolved Universal Terrestrial Radio Access (E-UTRA); Physical channels and modulation," TS 36.211, 3rd Generation Partnership Project (3GPP), Sept. 2008.

[6]   J.-J. van de Beek, O. Edfors, M. Sandell, S. K. Wilson, and P. O. Borjesson, "On channel estimation in OFDM systems," in Proc. IEEE45th Vehicular Technology Conf., Chicago, IL, Jul. 1995, pp. 815-819.

[7]   O. Edfors, M. Sandell, J.-J. van de Beek, S. K. Wilson and P. O. Borjesson, "OFDM channel estimation by singular value decomposition,"in Proc. IEEE 46th Vehicular Technology Conference, Atlanta, GA, USA,Apr. 1996, pp. 923-927.

[8]   S. D. Ma and T. S. Ng, "Time domain signal detection based on second-order statistics for MIMO-OFDM systems," IEEE Trans.Signal Process. vol. 55, no. 3, pp. 1150–1158, Mar. 2007.

[9]   S.D. Ma and T.S. Ng, "Semi-blind time-domain equalization for MIMO-OFDM systems", IEEE Transactions on Vehicular Technology, 57(4), 2219-2227, July, 2008.

[10] MM Rana"Channel estimation techniques and LTE Terminal implementation challenges" in Proc. International Conference on Computer and Information Technology pp. 545-549 December 2010

[11] M. Simko, D. Wu, C. Mehlführer, J. Eilert, D. Liu "Implementation Aspects of Channel Estimation for 3GPP LTE Terminals" in Proc. Proc. European Wireless 2011, Vienna, April, 2011.

11

PERFORMANCE ANALYSIS OF AMPLIFY-AND-FORWARD RELAY BASED COOPERATIVE SPECTRUM SENSING IN FADING CHANNELS

Shumon Alam, O. Olabiyi, O. Odejide, and A. Annamalai

Center of Excellence for Communication Systems Technology Research
Department of Electrical and Computer Engineering
Prairie View A & M University, TX 77446 United States of America
shalam2000@gmail.com, engr3os@gmail.com, femiodejide@yahoo.com
aaannamalai@pvamu.edu

ABSTRACT

Relay based cooperative spectrum sensing elevates the spectrum sensing problems in harsh propagation conditions if the relay nodes are positioned suitably for the dynamic spectrum access systems. This article investigates the performance of relay based spectrum sensing (energy detection based) in myriad fading environments using the area under the receiver operating characteristic curve (AUC) performance metric. AUC is a single statistical parameter, which provides better insight of detection probability compared to the receiver operating characteristic curve (ROC). The article includes the performance analysis of two-hop multi-relay amplify-and-forward (AF) relay network with channel state information (CSI) assisted relaying (variable relay gain) and blind (fixed) relay gain. A unified model is also investigated to describe the relay network for variable and fixed relay gain. Various fading and relay configurations are illustrated to show the efficacy of the energy detection based collaborative spectrum sensing in AF relay configuration using the AUC performance metric. Performance of the variable and fixed relay gain based AF relay network are analysed and compared considering different factors affecting energy sensing such as the per-hop signal-to-noise ratio (SNR), fading index, and the time-bandwidth product. The results can be readily used for the design of the cognitive relay network.

KEYWORDS

Area under Curve, Amplify and Forward, Cooperative Relay Network, Diversity Combining, Energy Detector, Spectrum Sensing

1. INTRODUCTION

Key to the emerging paradigm of opportunistic dynamic spectrum access (DSA) is the cognitive radios which can dynamically sense the environment and rapidly tune their transmission parameters to best utilize the vacant/underutilized spectrum bandwidth. The first key requirement for DSA system requires robust spectrum sensing capability in a manner that the DSA systems do not create interference for the incumbent users. Among the various known spectrum sensing techniques, blind sensing based on energy detection is perhaps the simplest and most versatile. However, the detection performance of energy detectors is severely limited by harsh propagation environments and becomes unreliable at low signal-to-noise ratio (SNR). This issue is further complicated by the noise uncertainty and shadowing problem.

A relay node between a primary signal emitter and a detector can be placed to increase the range of detection or to tackle the shadowing scenarios. Among all relaying protocol, amplify and forward (AF) or non-regenerative protocol is the simplest in which relay-nodes just amplify and forward received signal [1]. AF can be further divided into two groups, namely, i) CSI assisted/variable relay gain [2], in which relay node adapts the relay gain based on the CSI to cancel out the fading effect, and ii) blind/fixed gain relaying, in which relay gain is chosen arbitrarily or sometimes semi-blindly based on the CSI statistics [3]. Blind/fixed relaying does not require monitoring the CSI of the 1st hop at the relay and hence with the fixed gain, relay-node outputs signal with variable power. Systems with blind relaying mechanism in AF model are more attractive from the practical implementation perspective for its lower complexity and ease of deployment. In relay based cooperative spectrum sensing, the fusion center (FC) can make the decision of the presence of primary signal either based on the data fusion (soft) or decision (hard) fusion. The FC can take the advantages of receiver combining techniques such as the maximum ratio combining (MRC), square law combining (SLC) techniques to improve the performance of the spectrum sensing for data fusion. In decision fusion, local relay nodes take initial decisions based on the local observations and forward "1" or "0" to the FC for fusion over the narrow band control channel. "1" indicates the presence of the primary signal whereas "0" indicates the absence of the primary signal. The fusion can be done using the *k-out-of-N* system. This structure is a very popular type of redundancy in fault tolerant systems. The FC declares the presence of the active signal if at least *k-of-the-N* CRs detect the presence of the signal. Following the *k-out-of-N* rule, "OR", "AND", and "Majority" fusion rule can be found in literature (e.g., [1], [4] - [10]). Data fusion shows better spectrum sensing compared to decision fusion but decision fusion/hard fusion requires less communication overhead over the reporting channels (between CR and FC), which is attractive for practical implementation.

Recent works [1] and [5] have shown the performance of signal detection improves many times with the collaborative cooperation even at the low SNRs. But the existing relay based analytical frameworks have limitations for performance measurement in fading channels as well as in shadowing effects. Thus, there is a need to obtain a better analytical framework for analysing the energy-detection-performance using the relay concept. In [2] and [11], CSI assisted relaying protocol is utilized to determine the energy detector's performance but the development is not in generalized form. Atapattu [12] analysed energy detection performance for a fixed gain relay network over only Rayleigh channels using complementary AUC analysis, but the derivation is not in generalized form and the solution is intricate. In [13] and [14], detail performance studies of dual-hop end-to-end transmission system with fixed relay gain are found but not the energy detector's performance. The existing solutions mainly deal with the perfect *i.i.d.* Rayleigh channels, which is clearly not the case in practical scenarios. Also most studies are found to deal with network performance characterization and not in terms of the energy detector's performance.

Thus, motivated by the current limitations, we seek simple framework for analysing the CSS relay network in *i.i.d./i.n.d.* channel conditions subject to multipath fading and shadowing effects. We analyse energy detector's performance for the simple two-hop relay network using our prior developed unified analytical framework of energy detection using the AUC performance metric [15] . Since AUC is the portion of the area of a unit square, the value ranges from 0 to 1 but in practice, signal detection can be considered as the flipping of a coin. The random guessing produces the diagonal ROC line between $(0, 0)$ and $(1, 1)$ which has an area of 0.5. Thus, for practical detectors, the AUC value will not be less than 0.5 [16]. Thus, AUC is a good measure of detection probability; it represents the probability that choosing the correct decision at the detector is more likely than choosing the incorrect decision. We utilize the AUC metric to characterize the detection performance for both the variable and blind (fixed) AF relaying protocols. We develop generalized frameworks to characterize the AF relay based energy detection performance in myriad of fading channels. In particular, we illustrate examples with Nakagami-m and Rayleigh channels for variable and fixed relay gain network respectively. We also develop a unified solution using the AUC metric both for the variable and fixed relay gain system and include results over Nakagami-m fading statistics. We use the proposed frameworks subsequently to study the efficacies of diversity combining schemes such as the maximum ratio combining (MRC), square law combining (SLC) in multi-relay network.

To our best knowledge, energy detector's performance study for the AF relay network with the variable relay gain and unified model with the AUC measurement parameter is new while for fixed relay gain our framework with AUC metric is in generalized and simple form. The analysis could be readily used to choose right parameters for energy detectors in cognitive radio system. Our framework circumvents the need of sophisticated computer software to compute the numerical results. Further, the results of this article are more complete analyses of our initial findings that were published in [17] [18]. The major contributions are summarized as follows.

(i) A simple generalized framework is developed using AUC metric to analyse the performance of the energy detector in CSI assisted dual-hop multi-relay network, which can be used in myriad of fading environments.

(ii) A simple generalized framework over myriad of fading channels is developed to analyse the detector's performance in blind AF two-hop multi-relay network.

(iii) A unified model is developed to handle the variable/blind AF protocol for the spectrum sensing analysis.

The rest of the paper is organized as follows. Section 2 presents the system model for the AF relay based cooperative spectrum sensing using the energy detection scheme. Section 3 includes proposed framework of performance analysis with the AUC metric for the CSI assisted relay network. In Section 4, blind AF relay network is investigated. Section 5 includes the analysis of the unified performance model. Section 6 includes the numerical results while Section 7 includes the concluding remarks.

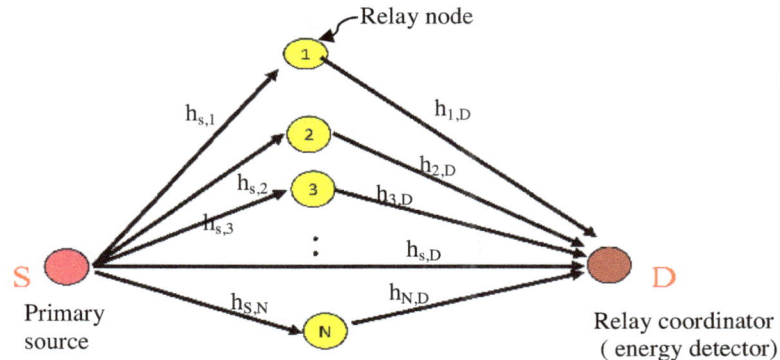

Figure 1. Relay based spectrum sensing

2. SYSTEM MODEL

Let us consider a two-hop N relay network (i = 1: N relay nodes) with variable/fixed relay gain as shown in Fig.1. S is the primary signal source and D is the relay coordinator. We also consider that the relay nodes amplify and forward the received noisy signal in orthogonal channels to the central relay coordinator for the soft / data fusion. Also a direct link is possible between S and D. $h_{S,D}$, $h_{S,i}$ and $h_{i,D}$ are the channel coefficients between source (S) & relay coordinator (D); S & i-th relays node; and i-th relay node & D respectively. Let us now consider a single relay system. The received primary signal at the i-th relay node and at the relay coordinator (D) can be given as

$$y_{S,i} = \sqrt{P_{S,i}} h_{S,i} \, x + n_{S,i} \qquad (1)$$

$$y_{S,D} = \sqrt{P_{S,D}} h_{S,D} \, x + n_{S,D} \qquad (2)$$

where, $P_{S,i} = P_{S,D}$ is the power of the transmitted source signal x while $n_{S,i}$ and $n_{S,D}$ are the additive white Gaussian noise (AWGN) introduced between the S and i-th relay node, and S and D. The relayed signal from the i-th relay to D is given by

$$y_{i,D} = G_i h_{i,D} y_{S,i} + n_{i,D} \qquad (3)$$

where, G_i is the amplifier gain of the i-th relay node and $n_{i,D}$ is the additive noise introduced between the i-th relay node and the relay coordinator (D). The gain can be variable or fixed gain.

We assume the detection at the relay coordinator can be described by the binary hypothesis (H_0, H_1) where H_0 means that the signal is absent whereas H_1 implies that the signal is present. The energy detection is performed by comparing the measured energy of the observed signal in the observation time interval T with the energy threshold λ [19]. The decision statistic (Y) is represented by the Chi-square distribution χ^2_{2u} under H_0 and a noncentral Chi-square distribution $\chi^2_{2u}(2\gamma)$ with $2u$ degree of freedom and noncentrality parameter of 2γ under hypothesis H_1 [19]. γ is the end-to-end instantaneous SNR and $u = TW$ is the time-bandwidth product (W is the filter bandwidth). Thus, the probability of the false alarm $P_f = \mathrm{Pr}\,ob\{Y > \lambda \,|\, H_0\}$ and the probability of the detection $P_d = \mathrm{Pr}\,ob\{Y > \lambda \,|\, H_1\}$ can be given as [20]

$$P_f(\lambda) = \Gamma(u, \frac{\lambda}{2}) / \Gamma(u) \qquad (4)$$

$$P_d(\gamma, \lambda) = Q_u(\sqrt{2\gamma}, \sqrt{\lambda}) \qquad (5)$$

where, $Q_u(.,.)$ is the generalized (u-th order) Marcum Q-function and $\Gamma(\cdot,\cdot)$ is the upper incomplete Gamma function. In terms of AUC metric, the average probability of false alarm and the detection probability can be given as [15] (see Appendix A for derivation)

$$\overline{A_{gen}}(\gamma) = \frac{1}{u2^u \Gamma(u)} \sum_{k=0}^{\infty} \frac{\Gamma(k+2u) \, {}_2F_1(1, k+2u; 1+u; 0.5)}{\Gamma(k+u) \, k! 2^{(u+k)}} (-1)^k \phi_\gamma^{(k)}(s)\,|_{s=1} \qquad (6)$$

$$\overline{A_{gen}}(\gamma) = 1 - \frac{1}{\Gamma(u)} \sum_{k=0}^{\infty} \frac{\Gamma(k+2u) \, {}_2F_1(1, k+2u; 1+u+k; 0.5)}{(u+k) \, \Gamma(k+u) \, k! 2^{(2u+k)}} (-1)^k \phi_\gamma^{(k)}(s)\,|_{s=1} \qquad (7)$$

where, $\phi_\gamma^{(k)}(s)$ is the k-th order derivative of the moment generating function (MGF) of the effective end-to-end signal-to-noise ratio (SNR). Equation (7) is an alternate solution to (6). These frameworks have advantage over any existing energy detection performance measuring framework since most common MGFs are available in literature [see 15, Table 1]. Also these frameworks can handle half odd integer value of u (bandwidth time product) and fading index m of Nakagami-m distribution as well as converge fast with nominal number of terms [15].

3. Relay Scheme with Variable Relay Gain

3.1. Single Relay with Two-hop network

Let us consider i-th relay network. Assuming flat fading and the variable relay gain $G_i^2 = 1/h_{S,i}^2$, the overall "exact" SNR of the harmonic means at the relay coordinator (D) can be given as [2]

$$\gamma_i \approx \frac{\gamma_{S,i}\,\gamma_{i,D}}{\gamma_{S,i} + \gamma_{i,D}} \tag{8}$$

where, $\gamma_{S,i} = (h_{S,i})^2 / N_0$ and $\gamma_{i,D} = (h_{ri,D})^2 / N_0$ are the SNR between the S and the i-th relay node and i-th relay to D respectively and N_0 is the one sided power spectral density (PSD) of the AWGN. If $\gamma_{S,D} = (h_{S,D})^2 / N_0$ is the SNR of this direct-path, between S and D, then the equivalent effective SNR at the relay coordinator with the direct-path can be given as

$$\gamma \approx \gamma_{S,D} + \gamma_i \tag{9}$$

Thus, the performance of the energy detection of the radio coordinator in the AF relay network with variable relay gain can be easily given by using the k-th order derivative of the MGF of the effective SNR (γ) in (6) or (7), which are in generalized forms.

The MGF of the two-hop i-th relay SNR at the receiver is available in literature. For instance, the MGF of tight bounded γ_i over Rayleigh $i.n.d.$ fading statistics is well-known, and is given as [2]

$$\phi_{\gamma_i}(s) = \frac{16}{3\Omega_{S,i}\Omega_{i,D}\left(A_1 + s\right)^2}$$
$$\left[\frac{4(1/\Omega_{S,i} + 1/\Omega_{i,D})}{(A_1 + s)}\,{}_2F_1\!\left(3,\frac{3}{2};\frac{5}{2};\frac{A_2 + s}{A_1 + s}\right) + {}_2F_1\!\left(2,\frac{1}{2};\frac{5}{2};\frac{A_2 + s}{A_1 + s}\right)\right] \tag{10}$$

where, $A_1 = \dfrac{1}{\Omega_{S,i}} + \dfrac{1}{\Omega_{i,D}} + \dfrac{2}{\sqrt{\Omega_{S,D}\,\Omega_{i,D}}}$, $A_2 = \dfrac{1}{\Omega_{S,i}} + \dfrac{1}{\Omega_{i,D}} - \dfrac{2}{\sqrt{\Omega_{S,D}\,\Omega_{i,D}}}$.

$\Omega_{S,i}$, $\Omega_{i,D}$ are the mean SNR between S & i-th relay-node and the i-th relay-node & D respectively and ${}_2F_1(.,.;.;.)$ is the Gaussian hypergeometric series. The simplified compact form of (10) for $i.i.d.$ case can be given as (11).

$$\phi_{\gamma_i}(s) = {}_2F_1\!\left(1,2;\frac{3}{2};\frac{-s\Omega_i}{4}\right) \tag{11}$$

where, Ω_i is the per hop i-th relay mean SNR. Also the MGF of γ_i of Nakagami-m channel with $i.i.d.$ fading statistics can be is given as [21] in (12).

$$\phi_{\gamma_i}(s) = {}_2F_1\!\left(m,2m;m+\frac{1}{2};\frac{-s\Omega_i}{4m}\right) \tag{12}$$

where, m is the Nakagami-m fading index. An alternative simple closed form of MGF of Rayleigh *i.n.d. cases* can also be found in [22, (55)] but (10) is more tractable for deriving the k-th derivative. The k-th derivative of (10) and (12) can be obtained easily and are given as (13) and (14) respectively by using the Leibnitz differential rule, identity [23, (0.430-1)] and

$$\frac{\partial^n}{\partial s^n}(A+s)^{-a} = -(1)^n (a)_n (A+s)^{-(a+n)}, \text{where}, (a)_n = a(a+1)...(a+n-1) \text{denotes}$$

the Pochhammer symbol.

$$\phi_{\gamma_i}^{(k)}(s)\Big|_{s=1} = \frac{64(\Omega_{S,i}+\Omega_{i,D})}{3(\Omega_{S,i}\Omega_{i,D})^2} \sum_{n=0}^{k}\binom{k}{n}(3)_{k-n}(A_1+1)^{-(3+k-n)}$$
$$\times \sum_{r=1}^{n}\frac{U_r}{r!}\frac{(3)_r(3/2)_r}{(5/2)_r}\,_2F_1\left(3+r,\frac{3}{2}+r;\frac{5}{2}+r;\frac{A_2+1}{A_1+1}\right)$$
$$+\frac{16}{3(\Omega_{S,i}\Omega_{i,D})}\sum_{n=0}^{k}\binom{k}{n}(2)_{k-n}(A_1+1)^{-(2+k-n)}$$
$$\sum_{r=1}^{n}\frac{U_r}{r!}\frac{(2)_r(1/2)_r}{(5/2)_r}\,_2F_1\left(2+r,\frac{1}{2}+r;\frac{5}{2}+r;\frac{A_2+1}{A_1+1}\right), \tag{13}$$

where,

$$U_r = \sum_{p=0}^{r-1}(-1)^p\binom{r}{p}\left(\frac{(A_2+1)}{(A_1+1)}\right)^p \sum_{\substack{m=0\\m\leq r-p}}^{n}\binom{n}{m}\frac{(r-p)!}{(r-p-m)!}$$
$$\times (A_2+1)^{(r-p-m)}(-1)^{n-m}(r-p)_{n-m}(A_1+1)^{(r-p+n-m)}$$

and

$$\phi_{\gamma_i}^{(k)}(s)\Big|_{s=1} = \left(\frac{-\Omega_i}{4m}\right)^k \frac{(m)_k(2m)_k}{(m+\frac{1}{2})_k}\times\,_2F_1\left(m+k,2m+k;m+\frac{1}{2}+k;\frac{-\Omega_i}{4m}\right). \tag{14}$$

Therefore, by using (13), and (14) in (6) or (7), AUC based summary detection performance of the coordinator for a single 2-hop relay system can be easily obtained for *i.n.d.* Rayleigh and *i.i.d.* Nakagami-m fading channels respectively. In case of *i.n.d.* Nakagami-m channels, the MGF of upper bounded $\widetilde{\gamma}_i$ ($\widetilde{\gamma}_i = \min[\gamma_{P,i},\gamma_{i,D}]$) can be given as [24, (11)] in (15)

$$\phi_{\widetilde{\gamma}_i}(s) = \sum_{\substack{k\in\{(P,i),(i,D)\}\\j\neq k}}\frac{\Gamma(m_k+m_j)}{\Gamma(m_k)\Gamma(m_j)}\left(\frac{\Omega_j m_k}{s\Omega_j\Omega_k+\Omega_j m_k+\Omega_k m_j}\right)^{m_k}$$
$$\times\frac{1}{m_k}\,_2F_1\left[1-m_j,m_k;1+m_k;\frac{(s\Omega_k+m_k)\Omega_j}{s\Omega_j\Omega_k+\Omega_j m_k+\Omega_k m_j}\right], \tag{15}$$

where, k is selected between primary source (*P*) and i-th relay node and j is selected between i-th relay and relay coordinator (*D*). We can easily get the k-th derivative of (15) by few algebraic steps or by using the Mathematica.

3.2. Multi-Relay with Two-hop network

Relay based cognitive/DSA radio network may have N relay systems and the radio coordinator can use various combining techniques such as maximum ratio combining (MRC) and square law combining (SLC) to coordinate the primary user's presence. It is assumed that, all the relay nodes forward and amplified primary user's signal through orthogonal channels such as the TDMA or FDMA techniques. Combining can be done either before (MRC method) or after detection (SLC method). Practical implementation of MRC is difficult to achieve as the coordinator requires the CSI information of each diversity branch in order to achieve coherent detection. This is not required in the case of SLC as the signals are non-coherently combined after detection; leading to twice the degree of freedom of the MRC; ultimately reduces the detection probability compared to MRC. For illustration, MRC and SLC are discussed below.

3.2.1 Maximal Ratio Combining (MRC)

MRC is a pre-combiner that linearly combines the signals from all diversity branches by first co-phased and weighted in proportional to their SNR. The approximated output SNR, γ_{MRC} of the MRC combiner for 2-hop N relay with the direct path can be given as [21]

$$\gamma_{MRC} \approx \gamma_{S,D} + \sum_{i=1}^{N} \frac{\gamma_{S,i}\gamma_{i,D}}{\gamma_{S,i}+\gamma_{i,D}} \tag{16}$$

The MRC decision statistics follows *i.i.d.* χ_{2u}^2 distribution under H_0 and $\chi_{2u}^2(\gamma_{MRC})$ under H_1 Hypothesis. Therefore, the P_f and the P_d at the MRC output for AWGN channels can be evaluated by (4) and (5) respectively. Thus, (9) is directly applicable to MRC technique for deriving the average AUC or the probability of detection by using the kth order derivative of the combined MGF of γ_{MRC}. The combined MGF is given as

$$\phi_{MRC}(s) = \phi_{S,D}(s)\prod_{i=1}^{N}\phi_i(s) \tag{17}$$

where, $\phi(s)_{S,D}$ and $\phi(s)_i$ are the MGFs due to the SNR of direct path and the i-th relay part respectively. Using the Leibnitz's differential product rule, the k-th derivative of (17) can be given as

$$
\begin{aligned}
\phi_{MRC(N+1)}^{(k)}(s) = \sum_{n_0=0}^{k}\sum_{n_1=0}^{n_0}\sum_{n_2=0}^{n_1}\cdots\sum_{n_{N-1}=0}^{n_{N-2}}\binom{k}{n_0}\binom{n_0}{n_1}\binom{n_1}{n_2}\cdots\binom{n_{N-2}}{n_{N-1}} \\
\times \phi_0^{(k-n_0)}(s)\big|_{s=1}\,\phi_1^{(n_0-n_1)}(s)\big|_{s=1}\,\phi_2^{(n_1-n_2)}(s)\big|_{s=1} \\
\times \cdots \phi_{N-1}^{(n_{N-2}-n_{N-1})}(s)\big|_{s=1}\,\phi_N^{(n_{N-1})}(s)\big|_{s=1}
\end{aligned}
\tag{18}
$$

where, $\phi_0(s) = \phi_{S,D}(s)$ and its k-th derivative over myriad fading statistics can be found in literature [see 15,TABLE-3-1]. $\phi_i^{(k)}(s)$ is the k-th derivative of the MGF $\phi_i(s)$ is due to the i-th relay parts and can be given either by (13) or (14). (N+1) in (18) indicates the N relays with the direct path between S and D. k-th derivative of (15) can also be used for representing *i.n.d.* Nakagami-m statistics. For illustration, considering the practical cases of $N = 2$ and 3 relays, (18) can be written as (19) and (20) respectively.

$$\phi_{MRC(2+1)}^{(k)}(s) = \sum_{n_0=0}^{k}\sum_{n_1=0}^{n_0} \binom{k}{n_0}\binom{n_0}{n_1}\phi_0^{(k-n_0)}(s)|_{s=1}\,\phi_1^{(n_0-n_1)}(s)|_{s=1}\,\phi_2^{(n_1)}(s)|_{s=1}. \qquad (19)$$

$$\phi_{MRC(3+1)}^{(k)}(s) = \sum_{n_0=0}^{k}\sum_{n_1=0}^{n_0}\sum_{n_2=0}^{n_1} \binom{k}{n_0}\binom{n_0}{n_1}\binom{n_1}{n_2}\phi_0^{(k-n_0)}(s)|_{s=1}$$
$$\times\,\phi_1^{(n_0-n_1)}(s)|_{s=1}\,\phi_2^{(n_1-n_2)}|_{s=1}(s)\phi_3^{(n_2)}(s)|_{s=1} \qquad (20)$$

Thus, using (19) and (20) in (6) or (7), average AUC measurement can be obtained for dual hop, 2 and 3 relays with direct path respectively over the Rayleigh or Nakagami-*m* statistics.

3.2.2 Square-Law Combining (SLC)

SLC is a post combiner technique that performs square-and-integrate operation over per branch-output and added to yield a new decision static Y_{SLC}. For N relays with one direct path, Y_{SLC} is the sum of $(N+1)$ i.i.d. χ_{2u}^2 for H_0 and $\chi_{2u}^2(\varepsilon_j)$ for H_1, where the noncentrality $\varepsilon_j = 2\gamma_{SLC}$, hence the P_f and the P_d can be expressed by (4) and (5) respectively by replacing u with $u(N+1)$. Therefore, for SLC scheme, (6) or (7) can be re-written by replacing u with $u(N+1)$. Then (19) and (20) can be used with the updated expressions of (6) or (7) to obtain the AUC analysis for SLC scheme for 2 and 3 relay networks.

4. RELAY SCHEME WITH BLIND (FIXED) RELAY GAIN

We consider blind/fixed relay gain in this section and follow the model description of Section 2. Let us consider for simplicity $\gamma_i\{i=1,2\}$ is the per-hop SNR for a dual hop relay network, where 1 indicates the path from S to i-th relay node and 2 indicates the path between i-th relay node and D. In [3], detail study of the dual-hop transmission with fixed relay gain can be found and for a single relay with two-hop network, the equivalent end-to-end SNR at the coordinator can be written as

$$\gamma_{eq} = \frac{\gamma_1\gamma_2}{\gamma_2 + C} \qquad (21)$$

where, C is a constant with fixed gain G such that $C = (\varepsilon_2 / G^2 N_{01})$ and ε_i is the source ($i = 1$) and relay ($i = 2$) transmitted signal power, N_{0i} is the AWGN signal power per hop. Thus, if we know the MGF of the equivalent SNR (21), we can find the k-th derivative for it. We can use this k-th derivative directly in (6) or (7) to obtain the average AUC.

For instance, if the two hops are assumed to suffer independent and not necessarily identically distributed Rayleigh fading, then γ_i is exponentially distributed with parameter $\bar{\gamma}_i = \varepsilon_i\Omega_i / N_{0i}$ ($i = 1, 2$), where $\Omega_i = h_i^2$ is the average fading power on the ith hop. Then, C and the MGF of the equivalent SNR over the Rayleigh fading channels (*i. n. d.*) with fixed relay gain is given in compact closed-form as [3]

$$C = \frac{\bar{\gamma}_1}{e^{1/\bar{\gamma}_1} E_1\left(\dfrac{1}{\bar{\gamma}_1}\right)} \qquad (22)$$

$$\phi_{\gamma_{eq}}(s) = \frac{1}{\bar{\gamma}_1 s + 1} + \frac{C\bar{\gamma}_1 s e^{\frac{C}{\bar{\gamma}_2(\bar{\gamma}_1 s+1)}}}{\bar{\gamma}_2(\bar{\gamma}_1 s + 1)^2} E_1\left(\frac{C}{\bar{\gamma}_2(\bar{\gamma}_1 s + 1)}\right) \qquad (23)$$

where, $E_1(.)$ in (22) is the exponential integral function defined as [25, (5.1.1)]. We use

$\dfrac{\partial^n}{\partial s^n}(A + s)^{-a} = -(1)^n (a)_n (A + s)^{-(a+n)}$ (where $(a)_n$ denotes Pochhamer symbol), identity [26,

(0.430-1)] and the Leibniz's differential rule, to obtain the kth order derivative of (23) as

$$\phi_{\gamma_{eq}}^{(k)}(s) = A + \sum_{n=1}^{k} \frac{U_n}{n!} F^{(n)}(y) \qquad (24)$$

where, $A = \dfrac{1}{\bar{\gamma}_1}(-1)^k k! (s + \dfrac{1}{\bar{\gamma}_1})^{-(k+1)}$, $y = \dfrac{C}{\bar{\gamma}_2(\bar{\gamma}_1 s + 1)}$

$$F^{(n)} = \sum_{p=0}^{n} \binom{n}{p} \left[ye^y (1 - y - 2n + 2p) + e^y(n-p)(2-n+p) \right] B, \quad B = (-1)^k E_{1-k}(y)$$

$$U_n = \sum_{r=0}^{n} (-1)^r \binom{n}{r} y^r \frac{d^k}{ds^k}(y^{n-r}),$$

$$\frac{d^k}{ds^k}(y^{n-r}) = (-1)^k \left(\frac{C}{\bar{\gamma}_1 \bar{\gamma}_2}\right)^{n-r} (n-r)_k (s + \frac{1}{\bar{\gamma}_1})^{-(n-r+k)}$$

Now substituting (24) into (6) or (7) and evaluating at $s = 1$, we obtain the average AUC for the two-hop relay network with fixed relay gain over the independent and not necessarily identically distributed Rayleigh channels.

Our two-hop single relay model can be easily extended to two-hop multi-relay system to utilize the diversity reception concept. Using the k-th-derivatives of the MGF of the received SNR for fixed relay gain with (18), we can simply derive coordinator's detection performance with fixed relay gain in *i.n.d.* fading environments for MRC or SLC scheme. Considering the relay part, for MRC and SLC, we can write following generalized schemes

$$\overline{A}_{gen}^{MRC}(\gamma) = 1 - \frac{1}{\Gamma(u)} \sum_{k=0}^{\infty} \frac{\Gamma(k+2u)\,_2F_1(1,k+2u;1+u+k;0.5)}{(u+k)\Gamma(k+u)k!2^{(2u+k)}}(-1)^k$$

$$\sum_{n_1=0}^{k}\sum_{n_2=0}^{n_1}\cdots\cdots\sum_{n_{N-1}=0}^{n_{N-2}} \binom{k}{n_1}\binom{n_1}{n_2}\binom{n_2}{n_3}\cdots\binom{n_{N-2}}{n_{N-1}} \qquad (25)$$

$$\times \phi_1^{(k-n_1)}(s)|_{s=1}\ \phi_2^{(n_1-n_2)}(s)|_{s=1}\ \cdots\cdots\phi_N^{(n_{N-1})}(s)|_{s=1}$$

$$
A_{gen}^{SLC}(\gamma) = 1 - \frac{1}{\Gamma(uN)} \sum_{k=0}^{\infty} \frac{\Gamma(k+2uN)\,_2F_1(1,k+2uN;1+uN+k;0.5)}{(uN+k)\Gamma(k+uN)k!2^{(2uN+k)}}(-1)^k
$$

$$
\times \frac{(-1)^k}{\ } \sum_{n_1=0}^{k} \sum_{n_2=0}^{n_1} \cdots \sum_{n_{N-1}=0}^{n_{N-2}} \binom{k}{n_1}\binom{n_1}{n_2}\binom{n_2}{n_3}\cdots\binom{n_{N-2}}{n_{N-1}}
$$

$$
\times \phi_1^{(k-n_1)}(s)\,|_{s=1}\ \phi_2^{(n_1-n_2)}(s)\,|_{s=1} \cdots \phi_N^{(n_{N-1})}(s)\,|_{s=1}
$$

(26)

where $\phi_{1...N}(s)$ are the MGFs of the $1...N$ relay parts and $\phi_i^{(k_1)}(s)$ is the k-th order derivative of $\phi_i(s)$.

5. UNIFIED PERFORMANCE ANALYSIS FOR THE AF RELAYING NETWORK

In this section we include the unified model to analyze energy detector's performance for the simple dual hop AF relay network. CSI assisted/variable relay gain and blind/fixed relay gain based system analysis are found in the literature and we also developed simple frameworks for variable and fixed relay gain using AUC metric in prior sections. Only limited studies are found regarding the unified model. In recent time, dual hop performance analysis based on the unified model can be found only in [13] and [14] but these did not analyze the energy detector's performance. Their frameworks are rather complicated and cannot be easily generalized. In this section we develop a simple unified model to analyze the energy detector's performance using our prior developed AUC approach. Our proposed detection performance analysis framework is capable of handling both the variable and fixed gain AF relay network. To our best knowledge, energy detector's performance study for the AF dual-hop relay network with the unified SNR model based on the AUC metric is new.

5.1 Unified Model Scheme

We consider the same relay network given by Fig. 1 but relay gain can be fixed or variable with per-hop SNR, γ_i. For dual-hop simple relay network, a general/unified model of the received SNR at the destination can be defined as [13]

$$
\gamma_{eq} = \frac{\gamma_1\gamma_2}{a\gamma_1 + \gamma_2 + b}
$$

(27)

where, a, b can take either 0 or 1. With the pair of a & b, special cases of AF relaying protocol can be defined. Variable gain and fixed gain relay configuration can be defined by (a, b) with (1, 0) and (0, 1) respectively. The relay gain (G_n) of nth relay can be determined as [13]

$$
G_n^2 = \frac{1}{ah_{1,n}^2} + bN_{0,i}
$$

(28)

where, $h_{1,n}^2$ is for the nth 1st hop relay. Thus, in general, deriving the MGF of (27) and taking its the higher order derivative and using it with (6) or (7), performance analysis can be obtained for the two-hop relay network either for CSI assisted or blind relay configuration where G can be set using (28).

In particular, we consider the two hops are assumed to suffer independent and not necessarily identically distributed Nakagami-m fading and thus γ_i is gamma distributed with parameter α_i and β_i ($i = 1, 2$). MGF of the unified SNR over the Nakagami-m channels (*i. n. d.*) can be found in [13]. As the special case when $a = 1$ and $b = 0$, the MGF ($\phi_{\gamma_{eq}}^{var}(s)$) based on the variable relay gain and when $a = 0$, $b = 1$, the MGF ($\phi_{\gamma_{eq}}^{fixed}(s)$), based on the fixed relay gain, can be given as (29) and (30) respectively.

$$\phi_{\gamma_{eq}}^{var}(s) = 1 - 2s \sum_{n=0}^{\alpha_1-1} \sum_{k=0}^{\alpha_2-1} \sum_{m=0}^{k} C_1(n,k,m) J_1(n,k,m) a^{\frac{n+m+1}{2}} \tag{29}$$

$$\text{where, } C_1(n,k,m) = \frac{a^{k-m} \beta_1^{\frac{n-m+1-2\alpha_1}{2}} \beta_2^{\frac{m-n-1-2k}{2}}}{m!(k-m)!n!(\alpha_1-n-1)!} \text{ ; and}$$

$$J_1(n,k,m) = \frac{\sqrt{\pi}\,\Gamma(\alpha_1+k+n-m+2)\Gamma(\alpha_1+k-n+m)}{\Gamma(\alpha_1+k+3/2)} \times \left(\frac{16a}{\beta_1\beta_2}\right)^{\frac{n-m+1}{2}}$$

$$\times \frac{{}_2F_1(\alpha_1+k+n-m+2, n-m+3/2; \alpha_1+k+3/2; \bar{s})}{\left(s+\left[\sqrt{\frac{1}{\beta_1}}+\sqrt{\frac{1}{\beta_2}}\right]^2\right)^{\alpha_1+k+n-m+2}}$$

$$\phi_{\gamma_{eq}}^{fixed}(s) = 1 - 2s \sum_{n=0}^{\alpha_1-1} \sum_{k=0}^{\alpha_2-1} \sum_{m=0}^{k} C_2(n,k) J_2(n,k) \tag{30}$$

$$\text{where, } C_2(n,k) = \frac{\beta_1^{\frac{n-m+1-2\alpha_1}{2}} \beta_2^{\frac{n+k+1}{2}}}{k!n!(\alpha_1-n-1)!} \text{, and}$$

$$J_2(n,k) = \frac{\Gamma(\alpha_1+1)\Gamma(\alpha_1+k-n)}{2\sqrt{b/(\beta_1\beta_2)}} (s+1/\beta_1)^{(n-k-2\alpha_1)/2}$$

$$\times \exp[b/(2\beta_2(1+\beta_1 s))] W_{\frac{n-2\alpha_1-k}{2}, \frac{n-k+1}{2}}(b/(\beta_2(1+\beta_1 s)))$$

where, $W_{\mu,\nu}(\cdot)$ denotes the Whittaker W function [26, (922)].

The k-th order derivative of (29) and (30) can be easily derived. Taking the kth order derivative of (29) and evaluating at $s = 1$ and inserting in (6) or (7) gives the average AUC for the variable relay gain configuration and similarly with the kth order derivative of (30) and evaluating at $s = 1$ and after insertion in (6) or (7) gives the average AUC, the detection performance measure, of the energy detector for the fixed relay gain based two-hop AF relay network. Also the k-th derivative of (29) and (30) can be used with (25) and (26) to obtain the analytical results for MRC and SLC combining schemes.

6. NUMERICAL RESULTS

This section includes the numerical results for various AF relay configurations. The numerical results are given in three sub-sections for CSI assisted, blind relay, and unified performance analysis.

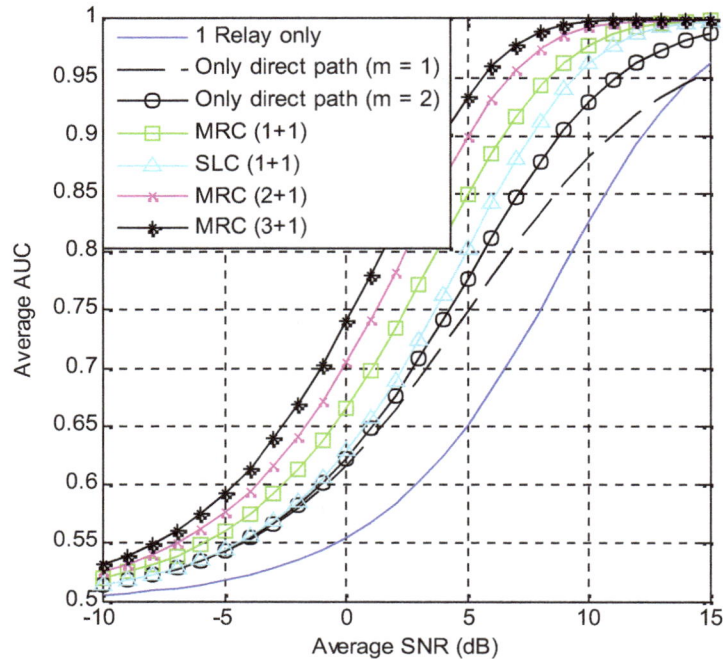

Figure 2. Average AUC vs. mean SNR for N-relay network with source to destination direct path over $i.i.d.$ Nakagami-m Channels ($m = 2$ and $u = 3.5$).

6.1 Numerical Results for CSI Assisted AF Relay Network

We have shown performance using the AUC metric using our simple yet useful generalized framework (6) or (7). Our generalized frameworks with appropriate higher order derivative of the equivalent MGF for the two-hop N relay model can be used over myriad of fading statistics. For illustration we have used Nakagami-m fading channels. We investigated relay based performance with or without direct path and also only direct path alone considering various parameters such as fading severity, number of relay paths, and time-bandwidth product. In particular, Fig.2 illustrates the performance of the detector for "only direct path", "only single relay", and "relay with direct path (N+1)" (N is the number of relays) cases. It is apparent that the performance of direct path (source to destination) is better than only the single relay configuration. Similar results were also found for the similar relay model for the symbol error rate in [27]. But the motivation of using relay network is to reduce the path loss when common receiver is at a disadvantage position such as the distance between transmitter and the receiver and at the hidden position/obstacle by large objects.

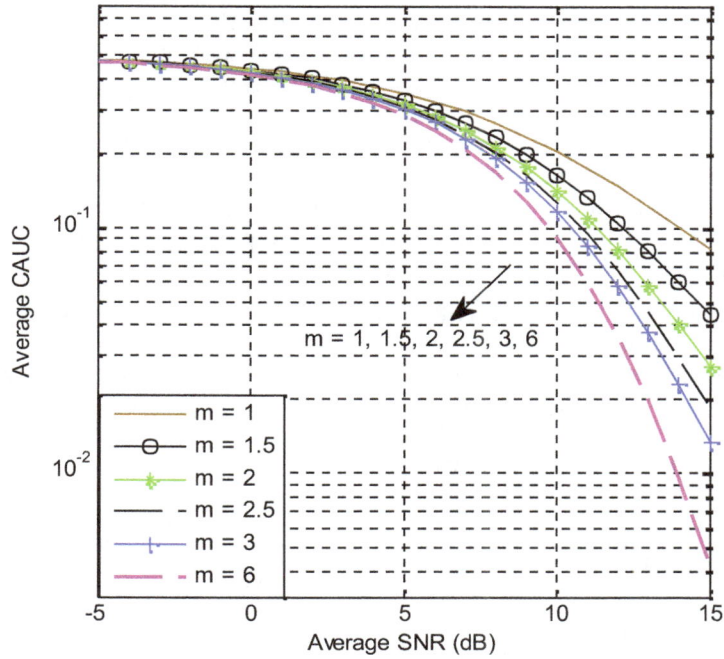

Figure 3. Average AUC vs. mean SNR for a relay network with source to destination direct path over *i.i.d.* Nakagami-m Channels, with m = 2 and u = 3.5.

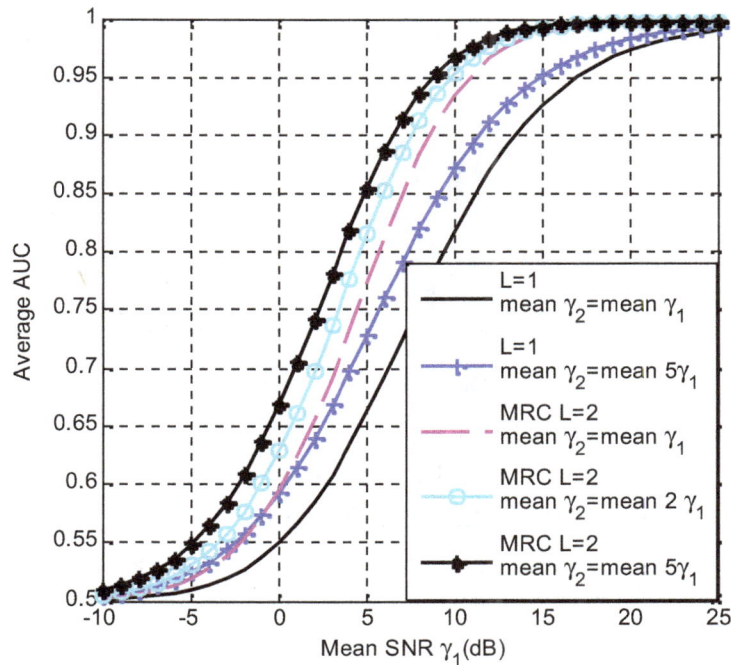

Figure 4. Average AUC vs. mean SNR for two hop fixed gain AF relaying over i.i.d. / i.n.d. Rayleigh fading with MRC illustration (u = 2).

Fig. 2 also illustrates the diversity combining results for MRC as well as for the SLC scheme. MRC shows the optimum performance as expected because of the higher end-to-end SNR associated with MRC. It is noticeable that with the diversity technique, detection capability improves greatly where direct-path gives a major impact on the detection performance. The figure also shows that the detection performance for "only-direct-path" outperforms the detection performance for "only single-relay" case even if the direct path's fading severity is higher. Overall, we see better performance with diversity combining compared to single relay reception at low SNR. The figure indicates that the relay based cooperative spectrum sensing can achieve a target robustness (say P_d=.5 and P_f=.1) even at low SNR.

In Fig. 3, the impact of the fading severity on the detection performance over *i.i.d.* Nakagami-m channels is illustrated. Only single relay is used to obtain the complementary AUC curves on the semi-log scale. From the figure it can be seen that the performance increases with order of m as m increases from 1 to 6. To show the advantages of our approach that it can handle the half odd-integer values of u and m, we also plotted AUC curves with half-odd integer u and half-odd integer m values in Fig.2 & 3 respectively.

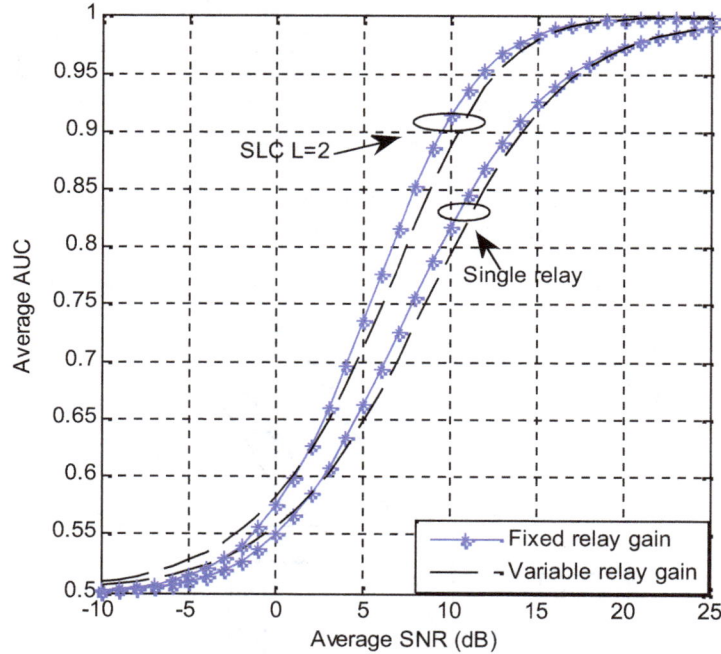

Figure 5. Comparison of fixed and variable relay-gain over i.i.d. Rayleigh channels for a dual-hop relay network ($u = 2$).

6.2 Numerical Results for Blind AF Relay Network

This section includes the performance of the relay coordinator based on the blind/fixed relay gain. Our result is much better than that of [12] since we have used the overall "exact" SNR of the harmonic means at the coordinator, whereas Atapattu [12] considered upper bounded SNR.

In Fig. 4, average AUC curves are plotted against the mean SNR for single relay and dual-relay MRC scheme for the blind AF dual hop relay network over the *i.n.d.* Rayleigh channels. For the second hop of each relay, SNR value is varied. With higher SNR value for the second hop, detector's performance improves considerably. A variable gain based average AUC for two-hop single relay and SLC (L = 2) scheme are derived and compared with the fixed gain based average AUC curve for *i.i.d.* Rayleigh fading in Fig. 5.

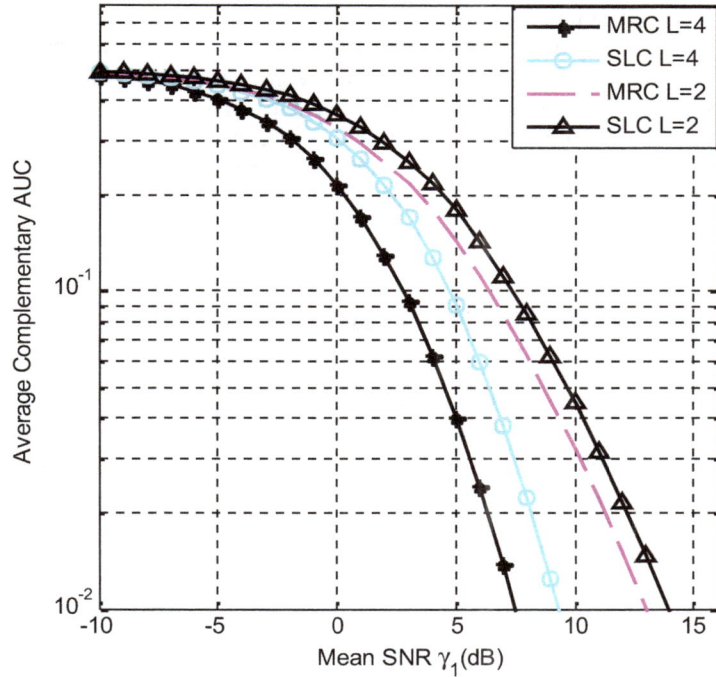

Figure 6. Complementary AUC. Comparison of MRC with SLC diversity combining schemes for a fixed AF dual-hop relay network over Rayleigh fading ($u = 2$, $\bar{\gamma}_2 = 5\bar{\gamma}_1$).

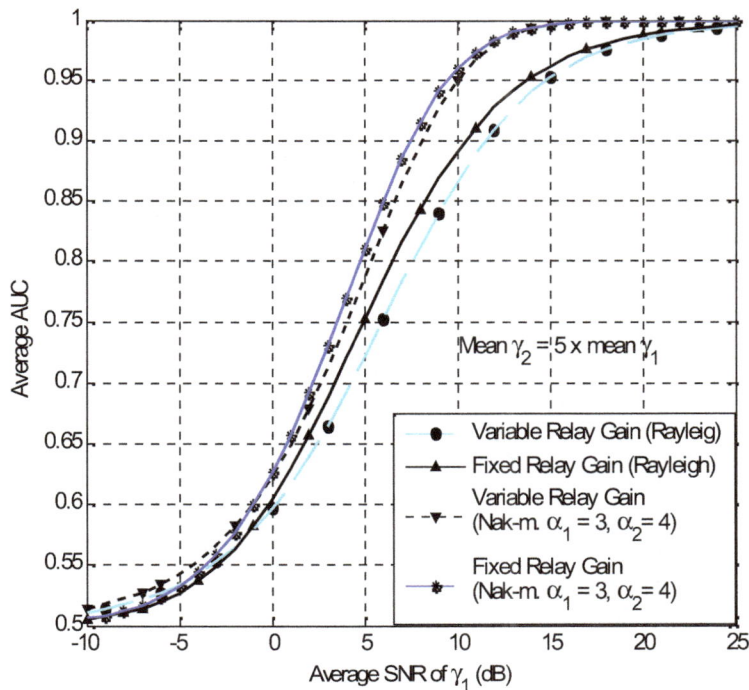

Figure 7. Fixed and variable relay gain performance over fading channels using unified SNR model with u = 2.

The result shows comparable performance and supports the results stated in [3]. Though, it is not shown here, we obtained that the increase of the time-bandwidth product (u) does not increase the AUC value for fixed relay number. Large sample (u) values do not necessarily increase the Average AUC; as u increases, the false alarm increases faster with larger samples, which is supported by [16]. Finally, in Fig. 6, average complementary AUC curves are shown on the semi-log plot for the MRC and SLC schemes for the fixed gain AF relay system over *i.n.d.* Rayleigh channels with $u = 2$. Clearly MRC outperforms SLC combining technique because of the higher end-to-end SNR associated with MRC. The curves confirmed that the higher number of relays (L) provide better detection performance.

6.3 Numerical Results for Unified AF Relay Network

This section includes the performance of relay coordinator for primary signal detection using the unified approach of equivalent SNR model. In Fig. 7, average AUC curves are plotted for a single relay system for the variable and fixed relay gain over the Rayleigh and Nakagami-m channels. The results shown are also supported by Fig. 5 and [3]. In Fig. 8, we compared variable and fixed relay gain based spectrum sensing performance for *i.n.d.* dual-hop single relay network. When first hop of the dual hop relay has significantly higher SNR than that of the second hop, it is observed that the detection probability is much higher for the fixed relay gain compared to the variable relay gain based configuration. Finally, in Fig. 9, using the unified model, average AUC curves of MRC and SLC schemes are compared for the dual-hop, dual-branch with direct path blind AF relay system over Nakagami-*m* channels with $\alpha_1 = 3$ and $\alpha_2 = 4$. Clearly, the curves follow the results obtained in Section 6.2.

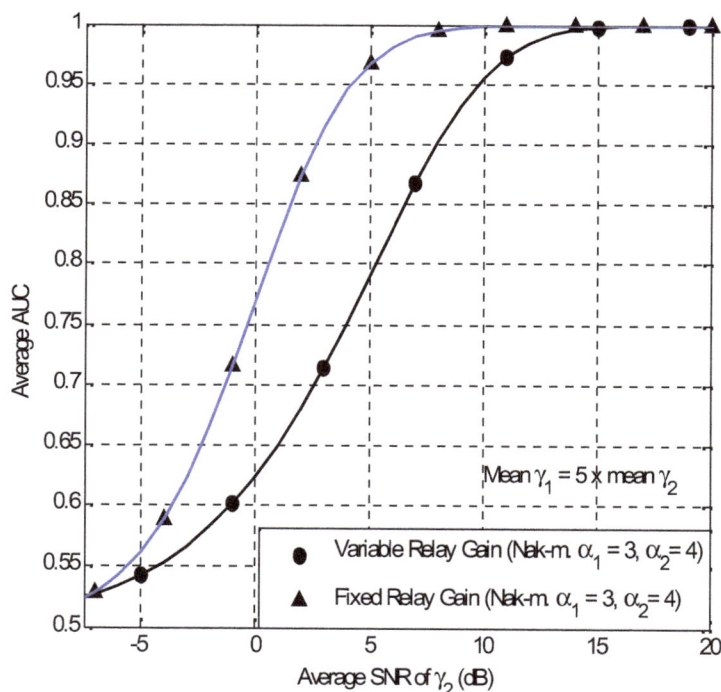

Figure 8. Comparison of variable relay gain and fixed relay gain based spectrum sensing

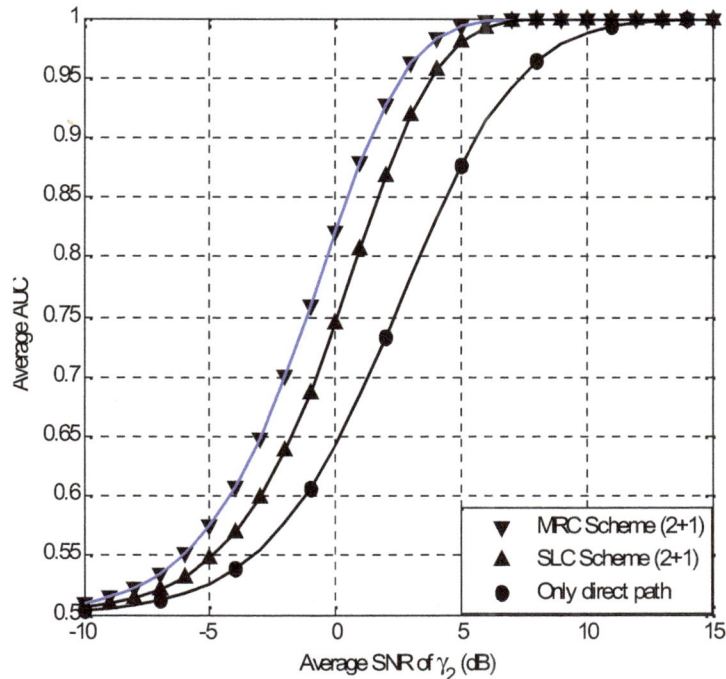

Figure 9. Comparison of MRC (N+1) and SLC (N+1) schemes, fixed gain AF dual-hop relay system over Nakagami-m channels with $u = 2$.

7. CONCLUSIONS

We have developed analytical frameworks to study the efficacies of cooperative spectrum sensing for the two-hop amplify-and-forward multi-relay network with variable and fixed-gain configurations. Even though we considered this as the relay coordinator's performance analysis, it is basically the concept of data fusion. Cooperative relaying techniques are of particular interest since such approaches could overcome the practical implementation issue of packing a large number of antenna elements on small-sized wireless devices while improving the detection performance by harnessing the benefits of distributed spatial diversity. We also developed two-hop unified model to analyze the fixed or variable gain based relay network. We have articulately shown the versatility of our frameworks for various scenarios. We considered practical parameters such as the fading severity and relay node number and illustrated that the receiver combining techniques with the relay network can achieve better detection probability and capable of satisfying target robustness.

The significant contribution of this work is the simplicity of the proposed method in analyzing the energy detector's performance in relay network. Also our method is capable of handling half odd-integer bandwidth-time product (u) and the half odd- integer Nakagami-m fading parameter (m). To the best of our knowledge, no prior work of CSI assisted AF relay network dealt with Nakagami-m channels. Also none of the prior work used AUC metrics except [12] but this cannot treat the half-integer u or m values. Also, their framework cannot be generalized for myriad of fading environments as well as their model only considered upper bounded equivalent SNR whereas our approach provides generalized frameworks with overall "exact" received SNR at the relay coordinator. This article has established the fast converging series-solutions of the energy detection performance using the MGF based frameworks for the two-hop AF relay network. The analytical results can be utilized readily for the practical implementation of the relay based cognitive radio network.

APPENDIX-A

With the alternative representation of the generalized Marcum Q-function [28], (5) can be written as

$$P_d(\gamma,\lambda) = Q_u\left(\sqrt{2\gamma},\sqrt{\lambda}\right) = \sum_{k=0}^{\infty} \frac{\gamma^k e^{-\gamma}}{k!} \frac{\Gamma\left(u+k,\frac{\lambda}{2}\right)}{\Gamma(u+k)} \tag{A-1}$$

Substituting $G(a,z) = \Gamma(a) - \Gamma(a,z)$ in (A-1) yields an alternate series as

$$P_d(\gamma,\lambda) = Q_u\left(\sqrt{2\gamma},\sqrt{\lambda}\right) = 1 - \sum_{k=0}^{\infty} \frac{\gamma^k e^{-\gamma}}{k!} \frac{G\left(u+k,\frac{\lambda}{2}\right)}{\Gamma(u+k)} \tag{A-2}$$

where, $G(a,z)$ is the lower incomplete Gamma function. Now $\overline{P}_d(\gamma,\lambda)$ can be obtained by averaging (A-1) over the PDF of channel SNR $f(\gamma)$ as

$$\overline{P}_{d-Gen} = \sum_{k=0}^{\infty} \frac{1}{k!} \frac{\Gamma\left(u+k,\frac{\lambda}{2}\right)}{\Gamma(u+k)} \int_0^{\infty} e^{-\gamma} \gamma^k f_\gamma(\gamma) d\gamma \tag{A-3}$$

Now, using following Laplace-transform identity, we obtain (6)

$$\int_0^{\infty} \gamma^k \exp(-\lambda\gamma) f_\gamma(\gamma) \, d\gamma = (-1)^k \left. \phi_\gamma^{(k)}(s)\right|_{s=\lambda} \tag{A-4}$$

Similarly, we can obtain (7) by averaging (A-2) and using the Laplace-transform (A-4).

REFERENCES

[1] K. Letaief and W. Zhang, "Cooperative communications for cognitive radio networks," *Proc. IEEE*, vol. 97, no. 5, pp. 878-893, May 2009.

[2] M. Hasna and M. Alouini, "End-to-end performance of transmission systems with relays over Rayleigh-fading channels," *IEEE Trans. Commun.*, vol. 2, pp. 1126–1131, Nov. 2003.

[3] M. Hasna and M. Alouini, "A performance study Of dual-hop transmissions with fixed gain relays," *IEEE Trans. Wireless Commun.*, vol. 3, no. 6, pp. 1163-1168, Nov. 2004.

[4] Z. Quan, S. Cui, and A. Sayed, "Optimal linear cooperation for spectrum sensing in cognitive radio networks," *IEEE J. Sel. Topics Signal Process.*, vol. 2, no. 1, pp. 28-40, Feb. 2008.

[5] G. Ganesan and Y. Li, "Cooperative spectrum sensing in cognitive radio Part I:Two user networks," *IEEE Trans. on Wireless Commun.*, vol. 6, no. 6, pp. 2204-2213, Jun. 2007.

[6] G. Taricco, "Optimization of linear cooperative spectrum sensing for cognitive radio networks," *IEEE J. Sel. Topics Signal Process.* , vol. 5, no. 1, pp. 77-86, Feb. 2011.

[7] E. Peh, Y. Liang, Y. Guan, and Y. Zeng, "Optimization of Cooperative Sensing in Cognitive Radio Networks: A Sensing-Throughput Tradeoff View," *IEEE Trans. Veh. Technol.*, vol. 58, no. 9, pp. 5294-5299, Nov. 2009.

[8] J. Shen, T. Jiang, S. Liu, and Z. Zhang, "Maximum channel throughput via cooperative spectrum sensing in cognitive radio networks," *IEEE Trans. Wireless Commun.*, vol. 8, no. 10, pp. 5166-5175, Oct. 2009.

[9] W. Zhang and K. Letaief, "Cooperative spectrum sensing with transmit and relay diversity in cognitive radio networks," *IEEE Trans. Wireless Commun.*, vol. 7, no. 12, pp. 4761-4766, Dec. 2008.

[10] R. Fan and H. Jiang, "Optimal multi-channel cooperative sensing in cognitive radio networks," *IEEE Trans. Wireless Commun.*, vol. 9, no. 3, pp. 2214-2222, Mar. 2010.

[11] S. Atapattu, C. Tellambura, and Jiang Hai, "Relay Based Cooperative Spectrum Sensing in Cognitive Radio Networks," in *Proc. IEEE GLOBECOM'09*, 2009, pp. 1-5.

[12] S. Atapattu, C. Tellambura, and H. Jiang, "Performance of Energy Detection: A Complementary AUC Approach," in *Proc. IEEE GLOBECOMC'10*, 2010, pp. 1-5.

[13] D. Senaratne and C. Tellambura, "Unified Exact Performance Analysis of Two-Hop Amplify-and-Forward Relaying in Nakagami Fading," *IEEE Trans. Veh. Tech.*, vol. 59, no. 3, pp. 1529-1534, Mar. 2010.

[14] M. Di Renzo, F. Graziosi, and Santucci, "A comprehensive framework for performance analysis of dual-hop cooperative wireless systems with fixed-gain relays over generalized fading channels," *IEEE Trans. Wireless Commun.*, vol. 8, no. 10, pp. 5060-5074, 2009.

[15] S. Alam, O. Olabiyi, O. Odejide, and A. Annamalai, "Simplified Performance Analysis of Energy Detectors Over Myriad Fading Channels: Area Under The ROC Curve Approach," *Int. J. Wireless Mobile Networks*, vol. 4, no. 4, pp. 33-52, Aug. 2012.

[16] S. Alam, O. Odejide, O. Olabiyi, and A. Annamalai, "Further Results on Area under the ROC Curve of Energy Detectors over Generalized Fading Channels," in *Proc. 34th IEEE Sarnoff Symp.*, NJ, 2011.

[17] S. Alam, O. Olabiyi, O. Odejide, and A. Annamalai, "Energy detector's performance evaluation in a relay-based cognitive radio network: Area under the ROC Curve (AUC) approach," in *Proc. IEEE GLOBECOM'11*, Houston, TX, DEC. 2011, pp. 338-342.

[18] S. Alam, O. Olabiyi, O. Odejide, and A. Annamalai, "A performance study of energy detection for dual-hop transmission with fixed gain relays: Area under the ROC Curve (AUC) approach," in *Proc. 22nd IEEE Personal Indoor Mobile Radio Commun.*, Toronto, Canada, Sep., 2011, pp. 1840-1844.

[19] H. Urkowitz, "Energy detection of Unknown deterministic Signals," *Proc. IEEE*, vol. 55, no. 4, pp. 523-531, April 1967.

[20] F. Digham, M. Alouini, and M. Simon, "On the energy detection of unknown signals over fading channels," in *Proc. IEEE Int. Conf. Commun.*, vol. 5, 2003, pp. 3575-3579.

[21] M.O. Hasna and M. S. Alouini, "Harmonic Mean and End-to-End performance of transmission system with realy," *IEEE Trans. Comm.*, vol. 52, no. 1, pp. 130-135, Jan. 2004.

[22] W. Su, A. K. Sadek, and K. J. Ray Liu, "Cooperative Communication Protocols in wireless networks:Analysis and optimum power allocation," *Springer Wwireless Personal and Communication*, vol. 44, pp. 181-217, 2007.

[23] I. Gradshteyn and I. Ryzhik, *Table of Integrals, Series and Products*, 7th ed. San Diego, CA: Academic, 2007.

[24] A. Annamalai, B. Modi, R. Palat, and J. Matyjas, "Tight Bounds on the ergodic capacity of cooperative analog relaying with adaptive source transmission techniques," in *Proc. IEEE PIMRC '10*, 2010, pp. 18-23.

[25] M. Abramowitz and I. A. Stegun, *Handbook of Mathematical Functions with Formulas, Graphs, and Mathematical Tables*. New York: Dover, 1970.

[26] I. Gradshteyn and I. Ryzhik, *Table of integrals, series and products*. San Diego, CA, USA: Academic, 2007.

[27] S. Haykin, D. Thomson, and J. Reed, "Spectrum sensing for cognitive radio," *Proc. IEEE*, vol. 97, no. 5, pp. 849-877, May 2009.

[28] A. Annamalai and C. Tellambura, "A simple exponential integral representation of the generalized Marcum Q-function QM (a, b) for real-order M with applications," in *Proc. 54th IEEE MILCOM*, 2008, pp. 1-7.

12

Study and Simulation of Quasi and Rotated Quasi Space Time Block Codes in MIMO systems using Dent Channel model

Priyanka Mishra[1],Rahul Vij[2],Gurpreet Singh[3] and Gaurav Chandil[4]

[1]Department of Electronics Engineering, United Group of Institution,Allahabad
mishrapriyanka6@gmail.com
[2]Department of Electronics Engineering, L R of Institute of Engineering,Solan
rahulvij2@gmail.com
[3]Department of Electronics Engineering, Shaheed Bhagat Singh State Technical Campus,
Ferozpur, Punjab
gurpreet2828@hotmail.com
[4]Department of Electronics Engineering, United Group of Institution,Allahabad
gaurav.iiitm18@gmail.com

ABSTRACT

Multiple Input Multiple Output (MIMO) has become one of the most exciting field in modern engineering.It has become one of the key technologies for wideband wireless communication systems.It is mainly used to increase capacity and data rate of any wireless systems.In this paper,we exploit the space and time diversity to decode the quasi and rotated quasi space time block codes (QOSTBC) based on dent channel model.For doppler shifting and rayleigh distribution we make use of dent channel model.A general Quasi and rotated quasi Space-Time Block Coded (STBC) MIMO structure is presented in this paper. BER analysis is presented in terms of code rate and diversity achieved using Quasi and Rotated quasi STBC methods. Furthermore, simulations are carried out using above two OSTBC methods with various modulation schemes in quasi static dent channel model and then optimum coding method is suggested for 4×2 quasi and rotated quasi Space time block coded MIMO systems.This provide fast decoding and gives better performance of communication system.BER analysis is presented in terms of diversity and code rate.

KEYWORDS

MIMO, QUASI ORTHOGONAL SPACE-TIME BLOCK CODES (QOSTBC), ROTATED QOSTBC,MAXIMUM LIKELIHOOD (ML) DECODING.

1. INTRODUCTION

MIMO technology constitutes a breakthrough in wireless communication system that offers a number of benefits that helps in improving the reliability of link. The advantage of MIMO technology includes improvement in array gain, spatial diversity gain, multiplexing gain and interference reduction. MIMO systems provide diversity to mitigate fading, that is realized by providing the receiver with multiple copies of the transmitted signal in space, frequency or time. By increasing number of signal replicas, the probability of getting least faded signal is increased. The information capacity of a system is increased by employing multiple transmit and receive antennas. An effective and practical way to gain the capacity of the multiple input multiple output (MIMO) is to employ Space Time (ST) Coding. Space-Time (ST) coding schemes combines coding along with transmit diversity to achieve high diversity performance. It can be implemented in two forms ST-Trellis and ST-Block codes.

The main problem with ST-Trellis scheme is that its decoding complexity increases exponentially with diversity and transmission rate. To address this problem Alamouti proposed Orthogonal ST block codes (OSTBC) for 2×1 and 2×2 systems.

Space-time block code designs have recently attractedconsiderable attentions. One attractive approach of space-timeblock codes (STBC) is from orthogonal designs as proposed by Alamouti [5] and Tarokh, Jafarkhani and Calderbank [9].STBC (Space-Time Block Code)[1][2][3] is a coding technique used in MIMO system which has time and space domain correlation among signals transmitted from the multiple antennae. Relative to the non coded system, STBC can provide higher gain of diversity and power with nobandwidth loss.OSTBC (Orthogonal Space-Time Block Code)[4][5][6] is one important subset of linear space time block coding, with its rows and columns keep orthogonal. One basic OSTBC is Alamouti code[7], which as standard, was adopted in third generation mobile telecommunication system[8]. OSTBC had one shortcoming, that is, when utilizing plurality modulation meanwhile the number of transmitting antenna is more than 2,even the OSTBC with full diversity gain obtained by using complex orthogonal, it still cannot obtain full transmissionrate.How to improve the transmission rate is one topic to study.Per above mentioned shorting, Jafarkhani proposed QOSTBC(Quasi-Orthogonal Space-Time Block Code).

These codes achieve full diversity and have fast MaximumLikelihood (ML) decoding at the receiver.It is a modulation schemes for multiple transmit antennas that provide full diversity with simple coding and decoding technique.Alamouti STBC scheme is the first space time block codes to provide full transmit diversity for two transmitting antenna. It is a simple single decoder scheme for 2×2 antennas that provide full rate. It is not possible provide full transmission rate for more than two antennas.Quasi orthogonal codes[8] of full rate have been proposed to overcome the shortcomings of orthogonal codes that cannot achieve full rate. In order to design full transmission rate that provide maximum possible diversity, the decoder performs pairwise symbol decoding instead of single symbol decoding. This is called quasi orthogonal space time block codes (QOSTBC) .Typically, quasi orthogonal space time codes perform best with ML decoding. This technique provide full rate with maximum possible diversity. It is impossible to achieve full diversity if all the symbols are chosen from the same constellation's ,the solution to this problem is rotation based method, which aims at maximizing the minimum distance in the space time constellation by using different constellation for different transmitted symbols. Using this concept it is possible to provide full diversity. This is called Rotated Quasi Orthogonal Space Time Block Codes.

The rest of the paper is organized as follows. In section 2, we introduce MIMO space time block code transceiver model and briefly review the ST code design criteria. In section 3,space time block codes is discussed with its two methods. In section 4, ML decoding method is discussed. The simulation results are presented in section 5, and some conclusions are drawn in section 6.

2. SYSTEM MODEL

A typical MIMO communication system consists of transmitter, channel and receiver. Space Time coding involves use of multiple transmit and receive antennas. Figure.1 shows the transceiver of MIMO in space time code. Bits entering to the system are mapped into the symbol mapper using different modulation techniques like BPSK, QPSK and 16-QAM.Bits entering the quasi and rotated quasi space time block code encoder serially are distributed to parallel substreams.Within each substream, bits are mapped to signal waveforms, which are then emitted from the antenna corresponding to that substream.Signals transmitted simultaneously over each antenna interfere with each other as they propagate through then wireless channel. The receiver collects the signal at the output of receiver antenna element and reverses the transmitter operation in order to decode the data with quasi and rotated quasi space time decoder.

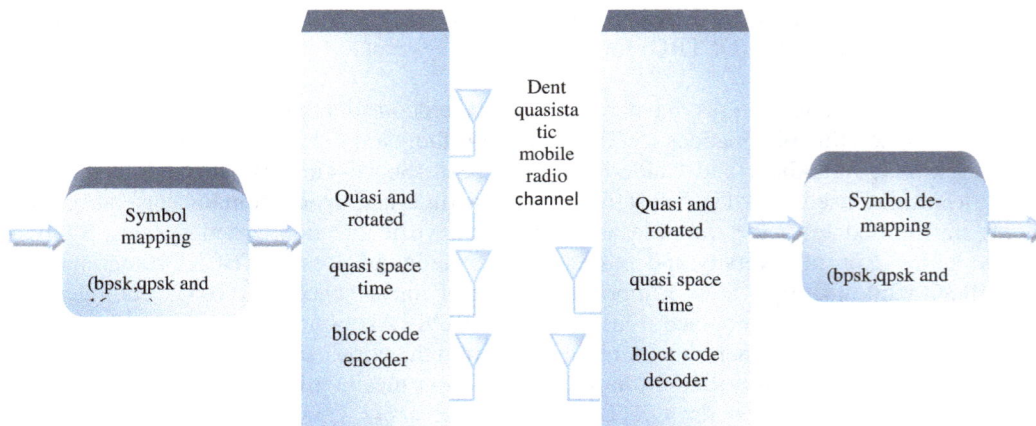

Figure 1. Quasi and rotated quasi Space Time Block Coded 4x2 MIMO transceiver structure

3. SPACE TIME BLOCK CODES (STBC)

Space time block codes(STBCs) [3] have been proposed to realize the enhanced reliability of multi-antenna systems.It is a transmit diversity scheme in which full diversity is achived while a very simple ML decoding algorithm is used at the decoder.This new paradigm uses the theory of orthogonal designs to design space time block codes [4].When transmitter has two antennas,Alaomuti codes [3] achieve the full diversity performance with a symbol rate of 1(rate-one) and simple linear processing under the assumption of no channel data information at the transmitter (CSIT) but perfect channel state information at the receiver (CSIR). Alamouti code provides the full diversity of 2 with 2 transmitting antenna with a rate of 1. For more reliable communication, Alamouti code can be further generalized for more than two transmitting antenna using the concept of orthogonal designs. But unfortunately it neither provide any coding gain nor achieve a rate larger than ¾ [3].

It is proved in [4] that a complex orthogonal design and corresponding Space Time Block code which provide full diversity and full transmission rate is not possible for more than two antennas.

3.1. QUASI ORTHOGONAL SPACE TIME BLOCK CODES (QOSTBC)

Full- rate orthogonal designs with complex elements in its transmission matrix are impossible for more than two transmit antennas[8].The only example of a full-rate full-diversity complex space-time block code using orthogonal designs is Alamouti schemes [8].The generator matrix [8] of Alamouti code is given as,

$$G\left(x_1, \quad x_2\right) = \begin{pmatrix} x_1 & x_2 \\ -x_2^* & x_1^* \end{pmatrix} \tag{1}$$

to design full rate codes, we consider codes with decoding pair of symbols [8].

$$G = \begin{pmatrix} G(x_1, x_2) & G(x_3, x_4) \\ -G^*(x_3, x_4) & G^*(x_1, x_2) \end{pmatrix} \tag{2}$$

$$= \begin{pmatrix} x_1 & x_2 & x_3 & x_4 \\ -x_2^* & x_1^* & -x_4^* & x_3^* \\ -x_3^* & -x_4^* & x_1^* & x_2^* \\ x_4 & -x_3 & -x_2 & x_1 \end{pmatrix} \qquad (3)$$

we denote the ith column of above matrix by v_i, then for any intermediate variable x_1, x_2, x_3, x_4, we have,

$$(v_1, v_2) = (v_1, v_3) = (v_2, v_4) = 0 \qquad (4)$$

where, the above symbols are inner product of eachother independently.

3.2. ROTATED QUASI SPACE TIME BLOCK CODES (RQOSTBC)

Sometimes, it is impossible to achieve code rate 1 for the complex orthogonal codes.To provide full diversity, different constellations are send through different transmitted symbols.This is done by rotating the symbols before transmission.This provide full-diversity with code rate 1 and these pairing of symbols gives good performance as compared to QOSTBC.

For M receive antennas, a diversity of $2M$ is achieved while the rate of the code is one. The maximum diversity of $4M$ for a rate one complex orthogonal code is impossible in this case if all symbols are chosen from the same constellation. By using same constellation for all symbols in the subset reduces the minimum distance for such codes. As a remedy to this problem, rotation based method is used that aims in maximizing the minimum distance in the space time constellation. To provide full diversity, we use different constellations for different transmitted symbols.

For example, we may rotate symbols x_3 and x_4 before transmission. Let us denote x'_3 and x'_4 as the rotated versions of x_3 and x_4, respectively. We show that it is possible to provide full-diversity QOSTBCs by replacing (x_3, x_4) with (x'_3, x'_4).The resulting code is very powerful since it provides full diversity, rate one, and simple pair wise decoding with good performance. Different modulation techniques use different rotation. In this paper, we are using bpsk, qpsk and 16-qam.The optimum rotation is given in table 1.

Table 1.Optimum rotation for different modulation techniques

Modulation techniques	Optimum rotation
BPSK	$\pi/2$
QPSK	$\pi/4$
16-QAM	$\pi/4$

3.3. DENT CHANNEL MODEL

A Rayleigh fading channel constitutes Dopplers spectrum is produced by synthesizing the complex sinusoids.The complex output of the jakes model [18], is given as,

$$h(t) = \frac{E_0}{\sqrt{2N_0 + 1}} \{ h_1(t) + jh_Q(t) \} \qquad (5)$$

The real and imaginary parts [7],is given as,

$$h_I(t) = 2\sum_{n=1}^{N_0}(\cos\varphi_n \cos v_n t) + \sqrt{2}\cos\varphi_N \cos v_d t$$

$$h_Q(t) = 2\sum_{n=1}^{N_0}(\sin\varphi_n \cos v_n t) + \sqrt{2}\sin\varphi_N \cos v_d t \qquad (6)$$

The unwanted correlation of Jake's model is removed in a modification by Dent model. The unwanted correlation canbe corrected by using orthogonal functions generated byWalsh-Hadamard codewords to weigh the oscillator valuesbefore summing so that each wave has equal power [11]. Theweighting is achieved by adjusting the Jake's model so thatthe incoming waves have slightly different arrival angles α_n .The modified Jakes model is given by

$$T(t) = \sqrt{(2/N_0)}\sum_{n=1}^{N_0}[\cos(\beta_n) + i\sin(\beta_n)]\cos(\omega_n t + \theta_n) \qquad (7)$$

where,the normalization factor $\sqrt{(2/N_0)}$ gives rise to $E\{T(t)T'(t)\} = 1$, $N_0 = N/4$,

$i = \sqrt{(-1)}$, $\beta_n = \Pi * n / N_0$ is phase,θ is initialphase that can be randomized to provide

different waveform realizations and $w_n = w_M \cos(\alpha_n)$ is the doppler shift. Dent's model successfully generates uncorrelated fadingwaveforms thereby simulating a Rayleigh multi-path air channel.

4. MAXIMUM LIKELIHOOD (ML) DECODING

At time T,four elements in the t_{th} rowof C are transmitted from the four transmit antenna.The codeword matrix is given as

$$C = G(s_1, s_2, s_3, s_4) \qquad (8)$$

Since,the four given symbols are transmitted in four time slots,this gives the code rate of 1.The ML decoding matrix for QOSTBC is given as,

$$\min_{s_1,s_2,s_3,s_4}\{H^H.C^H.C.H - H^H.C^H.r - r^H.C.H \qquad (9)$$

After simple calculations, ML decoding amounts to minimizing the given sum [4],

$$f_{14}(s_1,s_4) + f_{23}(s_2,s_3) \qquad (10)$$

where,

$$f_{14}(s_1,s_4) = \sum_{m=1}^{M}\left[\left(|s_1|^2 + |s_4|^2\right) + \left(\sum_{n=1}^{4}|\alpha_{n,m}|^2\right)\right.$$

$$+ 2R\left\{\left(-\alpha_{1,m}r_{1,m}^* - \alpha_{2,m}^*r_{2,m} - \alpha_{3,m}^*r_{3,m} - \alpha_{4,m}r_{4,m}^*\right)\right\}s_1$$

$$+\left(-\alpha_{4,m}r_{1,m}^{*}+\alpha_{3,m}^{*}r_{2,m}+\alpha_{2,m}^{*}r_{3,m}-\alpha_{1,m}r_{4,m}^{*}\right)s_{4}\right\}$$

$$+4R\left\{\alpha_{1,m}\alpha_{4,m}^{*}-\alpha_{2,m}^{*}\alpha_{3,m}\right\}R\left\{s_{1}s_{4}^{*}\right\}\right] \tag{11}$$

and,

$$f_{23}(s_{2},s_{3})=\sum_{m=1}^{M}\left[\left(|s_{2}|^{2}+|s_{3}|^{2}\right)\left(\sum_{n=1}^{4}|\alpha_{n,m}|^{2}\right)\right.$$

$$+2R\left\{\left(-\alpha_{2,m}r_{1,m}^{*}+\alpha_{1,m}^{*}r_{2,m}-\alpha_{4,m}^{*}r_{3,m}+\alpha_{3,m}r_{4,m}^{*}\right)s_{2}\right.$$

$$+\left(-\alpha_{3,m}r_{1,m}^{*}-\alpha_{4,m}^{*}r_{2,m}+\alpha_{1,m}^{*}r_{3,m}+\alpha_{2,m}r_{4,m}^{*}\right)s_{3}\right\}$$

$$+4R\left\{\alpha_{2,m}\alpha_{3,m}^{*}-\alpha_{1,m}^{*}\alpha_{4,m}\right\}R\left\{s_{2}s_{3}^{*}\right\}\right] \tag{12}$$

From, the above calculations symbols are independent and ML decoders decode the symbols separately.

5. SIMULATION RESULTS

In this paper,the simulation parameters used throughout in this work are listed out in Table 2. Results are then plotted and discussed using these simulation parameters.

5.1. SIMULATION PARAMETERS

Simulation parameters are shown MIMO space time block coding system given in figure 1. are listed in table 1.

Table 2.SIMULATION PARAMETERS FOR MIMO STBC

S.No.	Parameters	Values
1	No. of transmitters	4
2	No. of receivers	2
3	Max. Doppler shift(fm)	200Hz
4	Sampling frequency(fs)	8000Hz
5	Career modulation	BPSK,QPSK,16QAM
6	Bandwidth	20 MHz
7	Sampling time(ts)	1/fs
8	No. of Doppler shift(N)	8

5.2. RESULTS

The simulation parameters used throughout in this work are listed in Table 1 and 2.Results are then plotted and discussed using these simulation parameters.The simulation result is conducted in MATLAB.In this we will make comparision of performance of roatated QOSTBC with roatated QOSTBC using dent model.Results with different modulation techniques is plotted for

BER with SNR.The modulation technique used is BPSK,QPSK and 16-QAM with rotation of Π/2,Π/4 and Π/4 respectively for transmission of 1.5 bits/s/Hz.In figure.2 and figure.3,BER performance is better using BPSK as compared to QPSK and 16-QAM.Figure.4 shows the comparision of quasi and rotated quasi OSTBC with BPSK modulation,it is clearly observed that rotated quasi OSTBC gives better result when compared to quasi OSTBC.As it is clear from the figure by employing 4 transmitting antennas system performance is enhanced.

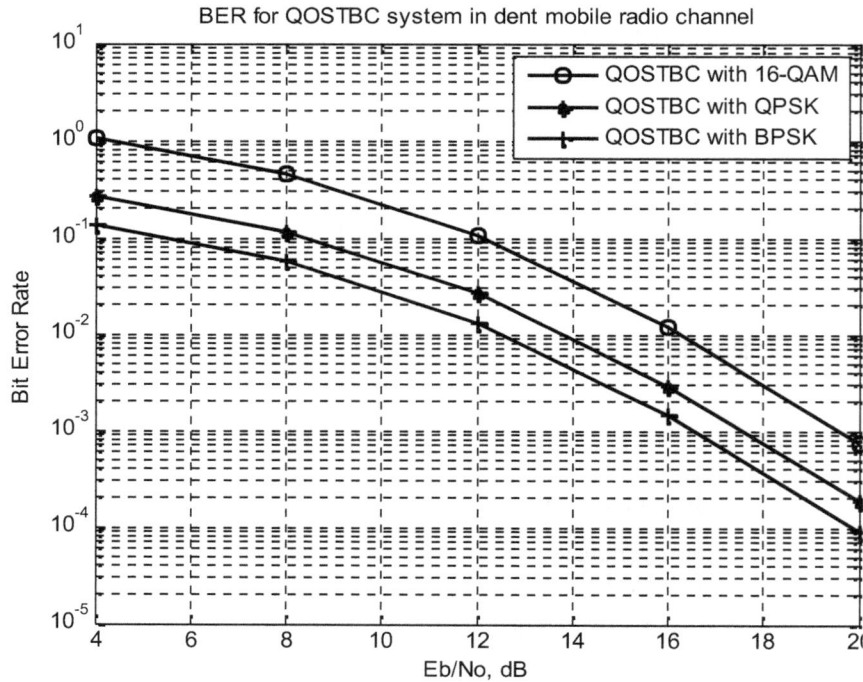

Figure 2.BER for Quasi-OSTBC system with dent channel model

6. CONCLUSION

In this paper, we studied MIMO system performance under mobile radio channel using dent model.Further,system performance is compared with three different modulation techniques and system with BPSK modulation gives better result as compared to oher modulation techniques.Quasi orthogonal space time block coding provide code rate of 1 and rotated quasi orthogonal space time block coding provide full rate and full diverty system with simple decoding technique.Maximum likelihood (ML) decoding reduces the decoding complexity of the system and enhances the system performance.it is clearly observed that the system performance enhances using dent channel model.

BER for rotated QOSTBC system in dent mobile radio channel

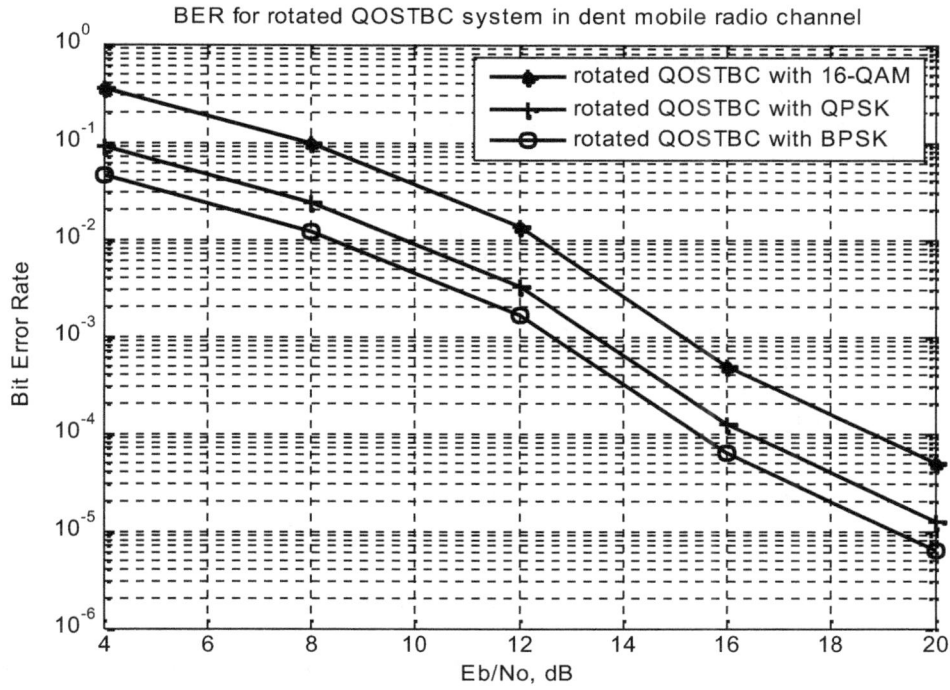

Figure 3.BER for rotaed quasi-OSTBC system with dent channel model

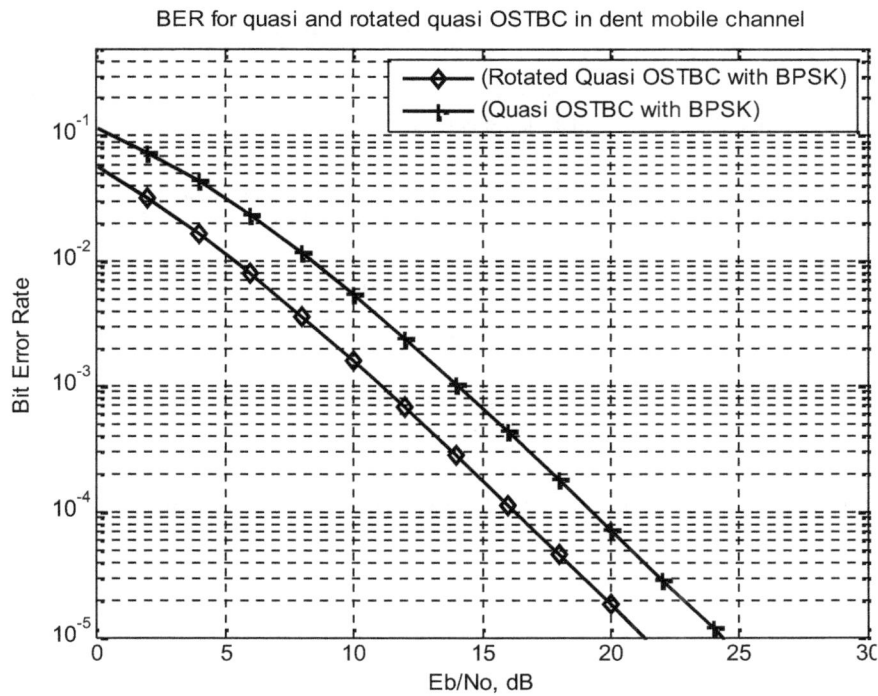

BER for quasi and rotated quasi OSTBC in dent mobile channel

Figure 4.BER for rotaed quasi and quasi OSTBC system with dent channel model

REFERENCES

[1]   A.F. Naguib, V.Tarokh, etc., "A Space-Time Coding Modem for High-Data-Rate Wireless Communications", IEEE Journal on Selected Areas in Communications, 1998, pp.1459-1477.

[2]   A. B. Gershman, N. D. Sidiropoulos, *Space-Time Processing for MIMO Communications*, John Wiley & Sons Ltd, 2005.

[3]   Mohinder Jankiraman, *Space-Time Codes and MIMO Systems*, Boston London, Artech House, 2004.

[4]   V. Tarokh, H.Jafarkhani and A R Calderbank, "Space-time block codes from orthogonal designs", IEEE Trans. Inform. Theory, vol.45, no.5,July 1999, pp.1456-1467.

[5]   Ganesan, G. and Stoica, P.(2001a). Space-time block codes: a maximum SNR approach, IEEE Transactions on Information Theory, 47(4), pp.1650-1656.

[6]   Wang H., Xia X.-G., "Upper bounds on complex orthogonal space-time block codes", In Proc. Of International Symposium on Information Theory, Lausanne, Switzerland. Also submitted to IEEE Transactions on Information Theory, 2002, pp. 303.

[7]   Flores, J.; Sánchez, J.; Jafarkhani, H.; , "Differential Quasi-Orthogonal Space-Frequency Trellis Codes," *IEEE Trans. Wireless Commun.*, vol.9, no.12, pp.3620-3624, Dec 2010.

[8]   Hamid Jafarkhani,"A Quasi-Orthogonal Space Time Block Codes", IEEE Transaction on Communications, vol.49, no.1, January 2001.

[9]   S.M. Alamouti,"A Simple Transmitter diversity scheme for wireless communications", IEEE J.Select. Areas Commun, vol.16,pp. 1451-1458,1998.

[10] Andreas A. Huttera, Selim Mekrazib, Beza N. Getuc and Fanny Platbrooda, "Alamouti-Based Space-Frequency Coding for OFDM," in Springer journal of wireless personal communications, vol. 35, no. 1-2, pp. 173-185, 2005.

[11] Yong Soo Cho,Jaekwon Kim,Won Young Yang,Chung Gu Kang,"MIMO-OFDM Wireless Communications with MATLAB", 1st edition, John Wiley & Sons (Asia) Pte Ltd, 2010.

[12] M. Rezk and B. Friedlander," On High Performance MIMO Communications with Imperfect Channel Knowledge " *IEEE Trans. Wireless Commun.*, vol. 10, no. 2, pp. 602-613, Feb. 2011.

[13] V.Tarokh, H.Jafarkhani and A.R. Calderbank,"Space Time Block Coding for wireless communications: Performance results", IEEE J. Select. Areas Commun, vol.17, pp. 451-460, Mar. 1999.

[14] A.V. Geramite and J. Seberry,Orthogonal Designs, Quadratic forms and Hadamard Matrices,Ser. Lecture Notes in Pure and Applied Mathematics.New York:Marcel Dekkar,1979,vol.43.

[15] G.J. Foschini and M.J. Gans,"On limits of wireless communications in a fading environment when using multiple antennas",Wireless Personal Communications, vol.6,pp.311-335,Mar. 1998.

[16] A.R. Hammons and H. El-Ganal,"On the theory of Space Time Codes for PSK modulation",IEEE Trans. Inform. Theory,vol.46,pp. 524-542,Mar. 2000.

[17] L. Zheng and D.Tse,"Diversity and multiplexing:a fundamental trade-off in multiple antennas channels",IEEE Transactions on Information Theory,vol.49,pp. 1073-1096,May 2003.

[18] W.C. Jakes,Ed.,"Microwave Mobile Communications",New York:Wiley,1974.

[19] G.L. Stuber,"Principles of Mobile Communication",Norwell,MA: Kluwer Academics Publishers,2nd ed.,2001.

[20] Bhasker gupta and Davinder S.Saini, "BER Analysis of Space-Frequency Block Coded MIMO-OFDM Systems Using Different Equalizers in Quasi-Static Mobile Radio Channel" *Proc. ofInternational Conference on Communication Systems and Network Technologies (CSNT-11)*, pp. 520-524, 3-5 June 2011.

[21] Wei Xiang, Julian Russell and Yafeng Wang,'" ICI Reduction Through Shaped OFDM in Coded MIMO-OFDM Systems", *International Journal on Advances in Telecommunications, vol 3 no 3 & 4, year 2010, http://www.iariajournals.org/telecommunications/*

13

Network Architecture And Performance Analysis Of Multi-OLT PON For FTTH And Wireless Sensor Networks

Monir Hossen and Masanori Hanawa

Interdisciplinary Graduate School of Medicine and Engineering
University of Yamanashi, Japan
mnr.hossen@gmail.com, hanawa@ieee.org

ABSTRACT

An integrated fiber-to-the-homes (FTTHs) and wireless sensor network (WSN) provides a cost-effective solution to build up an immaculate ubiquitous-City (U-city). The key objectives of effective convergence of FTTH and WSN are less computational complexity for data packet processing, low installation cost, and good quality of services. In this paper, we introduce an integrated network structure of multi-optical line terminal (multi-OLT) passive optical network (PON) which can accommodate multiple service providers in a single PON. A modified version of interleaved polling algorithm is proposed for scheduling of control messages from multiple OLTs in a single network. We also provide detailed numerical analysis of cycle time variation, successive grant scheduling time, and average packet delay for both uniform and non-uniform traffic loads generated by each ONU, using fixed service bandwidth allocation scheme and limited service bandwidth allocation scheme. We also compare the throughput of the proposed scheme with existing single-OLT PON for non-uniform traffic load using limited service bandwidth allocation scheme. The simulation results show that the proposed multi-OLT PON system can supports existing bandwidth allocation schemes with better performance than the single-OLT PON in terms of average packet delay, bandwidth utilization, and throughput.

KEYWORDS

Multi-OLT PON, Network Architecture, WSN, FTTH, Limited Service Bandwidth Allocation Scheme, Interleaved Polling Algorithm, Non-uniform Traffic, Packet Delay

1. INTRODUCTION

Currently, the network structure of a u-City is very complicated due to convergence of several new service providers. WSN is also one of the most important networks to build up a perfect u-City. Day-by-day the diameter of WSN becomes very large to provide the numerous new facilities and supports. Due to the large diameter as well as large number of wireless hops from the personal area network coordinator (PANC) to the surface nodes of a WSN, data packets from the surface nodes suffer from excess time delay. To solve this problem, clustering of large WSN is a good approach to reduce the individual network diameter as well as number of hops while the whole network size is enlarged. In the clustered WSN, every cluster consists of a cluster head (CH), and data from all nodes in a cluster are transmitted to a CH over a short distance with a small number of hops. So the data transmission delay in the clustered sensor network will be much lower than the case when each node directly communicates with the PANC through a large number of wireless hops. Even though, the main focuses of this paper is only on latency issue and analysis of energy efficiency is out of scope of this paper but the energy efficiency is another significant factor for an efficient WSN. Because the capacity of battery inside every sensor node and recurrent replacement of them is unrealistic that is why preservation of that energy is very important [1]. Some clustering algorithms of WSN have been developed to provide the energy efficiency. In [2], an energy efficient homogeneous clustering algorithm is proposed to extend the lifetime of sensor nodes and to maintain a balance energy consumption of nodes of a sensor networks.

Since in a u-city all the information systems are virtually linked together, it is very important to make a linkup with optical network terminals and WSNs. In [3, 4], a cluster based WSN is proposed where all sensor nodes of a cluster is connected with a CH through a small number of wireless hops and all the CHs are connected to a PANC by Radio-over- Fiber (RoF) links. Application of optical network in clustered WSN is very useful in that the optical attenuation is very small while wide bandwidth can be utilized.

PON is an optical network it does not contain active elements from source to destination which effectively reduces the active power consumption of the networks. PON provides several advantages over other access technologies. The main advantages of PON are high data rate, easy adaptability with new protocols and services, simple network structure, minimize the fiber deployment, less maintenance, and allow for long distance between the central office and customer premises. PON also provides effective solutions to satisfy the increasing capacity demand in the access part of the communication arena. In [5], a clustered WSN is proposed where all the CHs are connected with the PANC through a PON system. However, convergence of data networks and sensor networks in a single-OLT PON will increase the computational complexity for data packet processing in the OLT. To mitigate this problem some polling algorithms have been proposed to allow additional time in OLT for computation and management in addition with the guard time between every two successive ONUs [6]. Due to these increased computational complexity and additional computation time, some delay sensitive traffic of WSN will suffer from unexpected delay. To solve this problem, a single PON structure with multiple OLTs can be a good candidate.

In this paper, we propose a converged network structure of multi-OLT PON for FTTH terminals and clustered-based WSN. Here, we assume that a tree structured PON consists of two OLTs in the root side and N ONUs in the leave side sharing the same optical fiber links. One OLT is connected with all the FTTH terminals, and the other is connected with all the CHs of WSN. Although this system requires one additional OLT, it is inexpensive comparing with expenses for deploying two separate networks of data and WSN. We also propose a modified version of the interleaved polling algorithm [6], and scheduling algorithm for control messages for the multi-OLT PON system. The simulation results for successive grant scheduling time, evolution of cycle time, and average packet delay for both uniform and non-uniform traffic conditions are investigated and compared with a single-OLT PON for fixed service (FS) bandwidth allocation scheme and limited service (LS) bandwidth allocation scheme. Throughput of multi-OLT PON is also analysed and compared with the single-OLT PON for non-uniform traffic load considering LS bandwidth allocation scheme. Since generated packet sizes of FTTH terminals and CHs of WSN are not same, a comparative analysis is shown by changing the ratio of maximum bandwidth for FTTH and WSN because the maximum cycle time, T_{max}, is constant for PON system.

The rest of this paper is organised as follows. Section 2 presents the related works. In section 3, we presents the network architecture of the proposed multi-OLT PON. In section 4, we present a modified version of the interleaved polling algorithm, and a scheduling algorithm of a control message for multi-OLT PON. Section 5 shows the simulation results. Finally, section 6 draws conclusions.

2. RELATED WORKS

In this section, we describe some existing hybrid network architecture of WSN and FTTH terminals, interleaved polling algorithm, and bandwidth allocation algorithms for single-OLT PON system.

In the clustered WSN, usually all CHs are responsible to transmit data packet to the PANC through radio frequency (RF) transmission which suffers from data collision and more active energy consumption. However, conservation of energy in WSN is one of the prime requirements. To achieve this, J. Tang et al [3] proposed hybrid RoF based WSN architecture to combat the existing problems and to enhance the radio coverage and quality of service (QoS) for WSN. Here, whole sensor network is divided into several clusters and each cluster contains a static CH. All static CHs of WSN are distributed randomly in a wide area to collect the environment data such as temperature, humidity, the location of workers in the mine etc. Finally, all of the CHs are connected with the remote antenna units (RAUs) of RoF system to avoid data collision and RF transmission from CHs to PANC.

In [5, 7] a PON based WSN is proposed where synchronous latency secured (Sync-LS) medium access control (MAC) protocol and cooperative clustering algorithm are used to reduce the latency of a large sensor network. MAC protocols play a vital role for efficient data transmission and collision avoidance in wireless communication systems. Both of the papers provide latency efficiency in the wireless part of PON based WSN due to the cooperative clustering algorithm and Sync-LS MAC protocol using existing PON algorithms. But all these proposed algorithms are designed for single-OLT PON system where additional computational complexity of data packet processing in the OLT due to the convergence of data network terminals and WSN are not considered.

Interleaved polling algorithm [6] is a widely used algorithm in which OLT polls the ONUs individually and issues Grants to them in a round-robin fashion. In this algorithm OLT maintains a polling table containing the number of bytes waiting in each ONU's buffer and the round-trip-time (RTT) to each ONU. At the end of a transmission window, every ONU informs its queue size to the OLT by Report message. Interleaved polling algorithm has been developed to implement a dynamic bandwidth allocation scheme to improve the bandwidth utilization in PON system. The main principle of this algorithm is allocation of bandwidth to ONUs to avoid data collision and to fairly share the channel capacity. While this scheme is effective to avoid data collision, but in the single-OLT PON, it must requires some guard time between every two successive ONUs to avoid data overlapping.

In PON systems, each ONU's upstream bandwidth is decided by allocated time slots specified by the OLT in unit time [8]. Bandwidth allocation algorithm has a major impact on minimizing latency, improving fairness, meeting quality of services guaranties, and requirement of buffer size in upstream direction. In general sense, bandwidth allocation algorithms can be classified into two major groups; fixed bandwidth allocation and dynamic bandwidth allocation (DBA) algorithms.

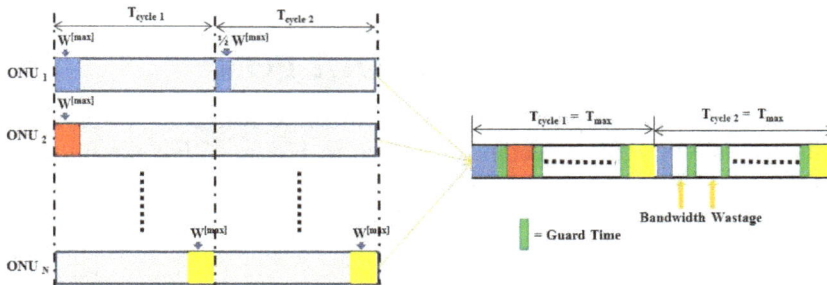

Figure 1. Fixed bandwidth allocation scheme

In [9], the performance of EPON using a fixed bandwidth allocation algorithm is studied where all traffics are considered to a single class. This scheme is simple and always grants the maximum window to all ONUs, as a result the cycle time T_{cycle} is constant for all traffic loads as shown in Figure 1. The main drawback of this algorithm is light-loaded ONUs will under-utilize their allocated bandwidth leading to increased delay to other ONUs and eventually deteriorate the throughput and bandwidth utilization of the system.

DBA schemes provide flexible bandwidth sharing of allocation among users. DBA is suitable for burst traffic such as FTTHs and VoIP. Different DBA algorithms have been developed to improve the bandwidth utilization and to adopt with the current demand of huge traffic. In the LS scheme [6, 10], the time-slot length of each ONU is upper bounded by the maximum bandwidth, $W^{[max]}$. When the requested bandwidth by the ONU is less than $W^{[max]}$, the OLT grants the requested bandwidth; otherwise, $W^{[max]}$ is granted. Figure 2 shows the LS bandwidth allocation scheme. Since the granted window is based on the requested window, the cycle time T_{cycle} is variable. As shown in the figure, cycle time for first cycle is $T_{cycle\ 1}= T_{max}$ because every ONU requests for maximum bandwidth $W^{[max]}$. On the other hand, cycle time in the 2nd cycle is $T_{cycle\ 2}= T_{max}- T_S$, here, T_S is the cycle time saving due to light-loaded ONUs. This scheme reduces the bandwidth wastage by granting smaller bandwidth to the light-loaded ONUs. However, one limitation of this algorithm is that making T_{cycle} too small will result lower bandwidth utilization because of constant guard time for every two successive ONUs.

In [11], a new DBA is proposed where ONUs in the network are divided into two sets, one set contains the ONUs with bandwidth guaranteed services while the second set contains the ONUs with best effort services. This scheme is able to provide guaranteed bandwidth for premier subscribers while best effort services are provided to other subscribers. However, all of these DBA algorithms are proposed only for the single-OLT PON and must require guard time between every two successive ONUs to avoid data overlapping. Due to this guard time, some bandwidth wastage problem is observed.

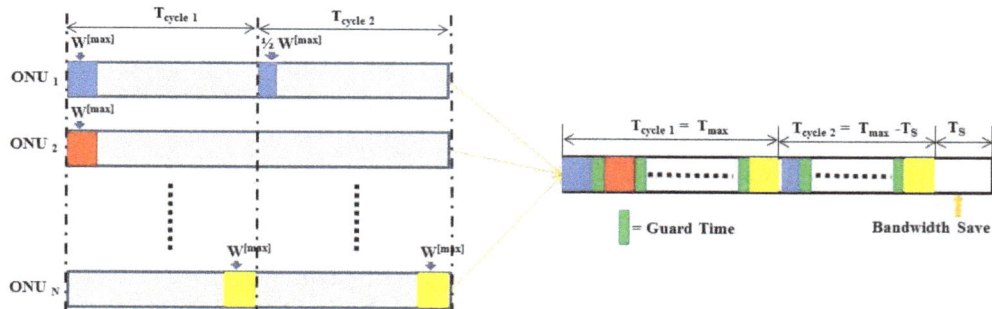

Figure 2. Dynamic bandwidth allocation scheme

3. NETWORK ARCHITECTURE OF MULTI-OLT PON

In this paper, we consider an access optical network architecture consisting of two OLTs and N ONUs connected using a single PON structure as shown in Figure 3. For simplicity, only two OLTs, OLT1 and OLT2/PANC, and four ONUs with tree based network topology are shown in the figure to explain the network structure and data transmission sequences for both upstream and downstream.

All transmissions in the proposed multi-OLT PON are performed between two OLTs in the root side and four ONUs in the leaf side of the tree topology. Here, OLT1 is connected with FTTH data terminals, ONU1 and ONU3. On the other hand, OLT2/PANC is connected with the static CHs of WSN, ONU2 and ONU4. All the connections between OLTs and ONUs are established through optical fibers and a passive splitter/combiner.

In downstream transmission, both OLTs follow the same polling table to initiate a transmission of a Grant message to the ONUs. Depending on the RTT delay between OLTs and ONUs, the 1st Grant message can be scheduled by any of both OLTs, because RTT depends on the physical distances between OLTs and ONUs and these distances are not fixed for all ONUs. Since, all downstream transmissions are broadcasted (point to multi-point transmission) from OLT to ONUs, both OLTs broadcast their Grant messages through an optical splitter and each ONU filters the received packets according to its destination address.

In upstream transmission, all ONUs share a common channel capacity and resources, and various multiple access schemes are exist to share a common channel in a PON-based access network. In our model, we consider time-division multiple access (TDMA) scheme to ensure the use of a single upstream wavelength for all users and a single receiver in the head end to reduce the system cost. According to the principle of the proposed multi-OLT PON system, OLT1 accepts data only from the ONUs of FTTH terminals (ONU1, ONU3) and OLT2 accepts data only from the ONUs of WSN (ONU2, ONU4).

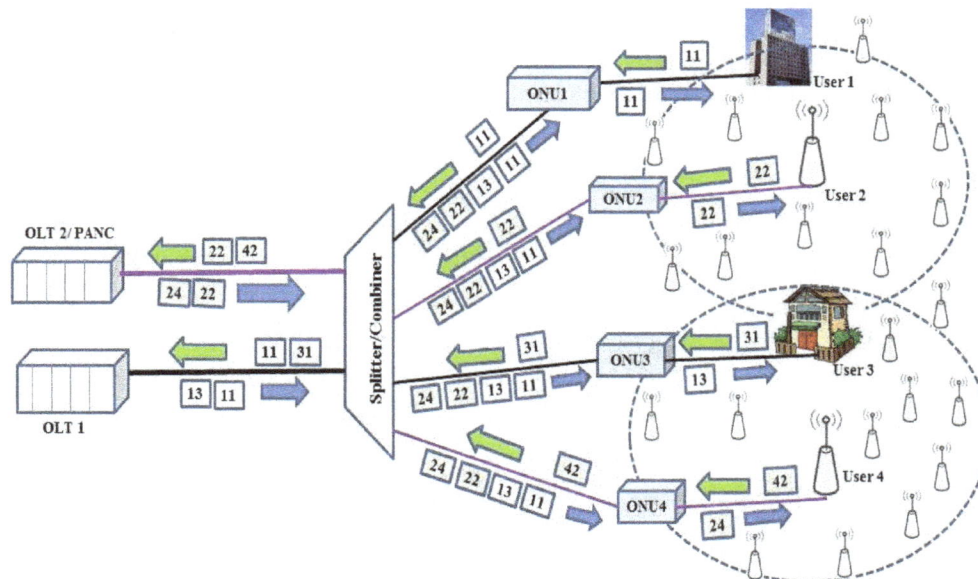

Figure 3. Network architecture and data transmission in proposed multi-OLT PON

4. MODIFIED ALGORITHMS FOR MULTI-OLT PON

In this section, the modified version of interleaved polling algorithm and scheduling algorithm of a control message suitable for the multi-OLT PON system are described.

4.1. Interleaved polling algorithm

The main aspect of polling algorithm is the scheduling of data transmission of ONUs. Usually, a polling protocol is cycle based, which used to avoid high traffic load as well as data collision, and it limits the maximum transmission window for each ONU. A commonly used polling algorithm is round-robin, which orders the transmission of every ONU periodically. To improve the network performance, interleaved polling with adaptive cycle time (IPACT) [6] is used. Interleaved polling algorithm can also have different policies, such as interleaved polling algorithm with and without stop polling. In all interleaved polling algorithms, an OLT contains a polling table which provides information about the RTT to every ONU and their granted window size. The knowledge of RTT and granted window size from the polling table are used to avoid data collision and overlapping of packets from different ONUs. Additionally, a guard time is used between every two successive ONUs to avoid overlapping of transmission windows by fluctuations in RTT of different ONUs [12] and on/off timing of laser of OLT and ONUs. Without using this guard time a single-OLT PON cannot ensure the overlap free transmission between two successive ONUs.

The data transmission from multiple OLTs and ONUs also must be well scheduled to improve network performance and to avoid data collision. A modified version of the interleaved polling algorithm [6] is used for the proposed multi-OLT PON system, where a common polling table is considered for both of the OLTs. Figure 4 shows the modified interleaved polling algorithm and RTT calculation procedure for the proposed multi-OLT PON system. For simplicity, ONU 1, 3, 5 ... (2i-1) (i is integer) are for the FTTH terminals and those are connected with OLT1. On the other hand, ONU 2, 4, 6 ... 2i are for the CHs of WSN and those are connected with OLT2.

In the multi-OLT PON, no guard time is required, because data of every two successive ONUs will be received by two different OLTs. So there is no possibility of data overlapping due to fluctuations of laser on/off timing and RTT. After receiving of data from a particular ONU, every OLT gets enough time before receiving data from the next ONU. This way, packet delay of the network and computational complexity of OLTs can be decreased while bandwidth utilization will be increased.

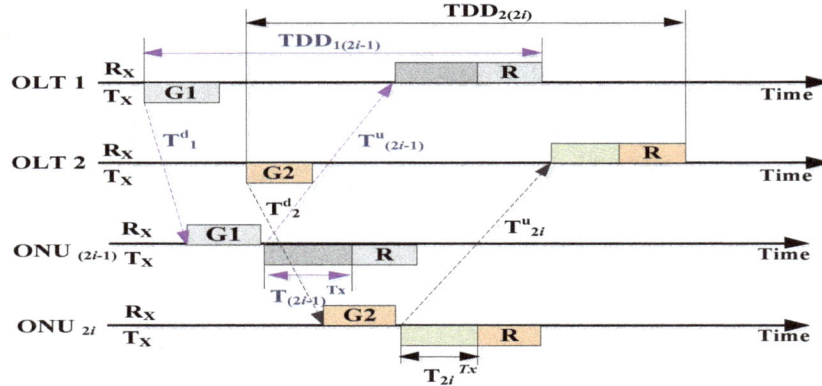

Figure 4. RTT and data transmission time in multi-OLT PON

Equations (1) and (2) represent the principle of RTT calculation for OLT1 and OLT2, equations (3) and (4) represent data transmission delay for OLT1 and OLT2, respectively.

$$RTT_{1(2i-1)} = T_1^d + T_{(2i-1)}^u \tag{1}$$

$$RTT_{2(2i)} = T_2^d + T_{2i}^U \tag{2}$$

$$TDD_{1(2i-1)} = T_G + T_1^d + T_{(2i-1)}^{Tx} + T_{(2i-1)}^u + T_R \tag{3}$$

$$TDD_{2(2i)} = T_G + T_2^d + T_{2i}^{Tx} + T_{2i}^u + T_R \tag{4}$$

where, T_G and T_R are the transmission times of Grant and Report messages, T^d is the downstream propagation delay, T^{Tx} is the upstream transmission time of data, and T^u is the upstream propagation delay.

Transmission time and propagation delay of data depend on the data transmission speed of the PON and physical distance between OLTs and ONUs. Usually, physical distance between an

OLT and ONUs are not equal [13] but the data transmission speed is a constant for TDMA PON. In TDMA PON, downstream traffic is broadcasting in nature which is handled by OLTs and upstream traffic from ONUs is allowed at a particular transmission time [14]. In our analysis, random distances between OLTs and ONUs but constant transmission speed are considered.

4.2. Scheduling algorithm of a control message

During upstream transmission in a single-OLT PON, all ONUs share a single uplink optical fiber trunk connected with an OLT. To prevent data collision due to multiple ONUs transmitting at the same time, a multi-point control protocol (MPCP) is being developed [13]. Usually, the MPCP operation in PON requires two control messages, Grant and Report. A Report message contains current queue length of each ONU to inform the OLT. On the other hand, the OLT assigns a timeslot to an ONU by a Grant message. Scheduling of Grant messages from OLT depends on RTT and granted transmission window size of ONUs [6]. This is because of the variation of RTT and granted window size for different ONUs.

Figure 5 shows a scheduling diagram of control messages for the proposed multi-OLT PON system. As the scheduling of the Grant messages depend on the RTT and granted window size of different ONUs, the starting Grant message can be sent by any of the both OLTs.

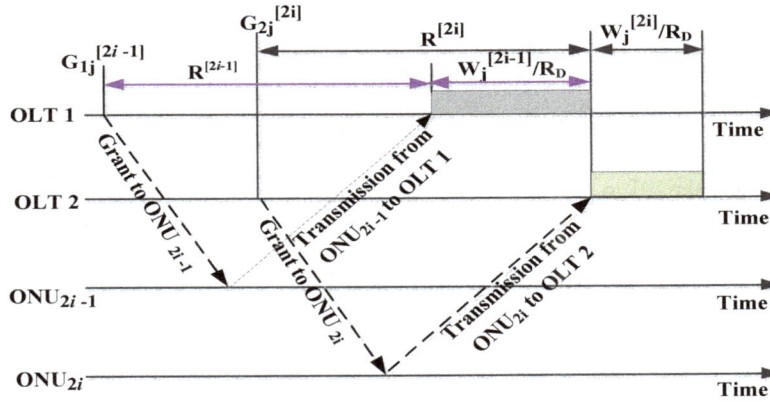

Figure 5. Scheduling diagram of control messages in multi-OLT PON

The scheduling of Grant messages as shown in above figure can be expressed as the following equations:

$$G_{2j}^{[2i]} = G_{1j}^{[2i-1]} + R^{[2i-1]} - R^{[2i]} + \frac{W_j^{[2i-1]}}{R_D} \tag{5}$$

$$G_{1j}^{[2i+1]} = G_{2j}^{[2i]} + R^{[2i]} - R^{[2i+1]} + \frac{W_j^{[2i]}}{R_D} \tag{6}$$

where, $G_{2j}^{[2i]}$ and $G_{1j}^{[2i+1]}$ are the time epochs for OLT2 and OLT1 when j^{th} Grant to ONU$_{2i}$ and ONU$_{2i+1}$ are transmitted respectively, $G_{1j}^{[2i-1]}$ is the time epoch for OLT1 when j^{th} Grant to ONU$_{2i-1}$ is transmitted, $R^{[2i-1]}$ and R$^{[2i]}$ are the RTT from OLT1 to ONU$_{2i-1}$ and OLT2 to ONU$_{2i}$ respectively, $R^{[2i+1]}$ is the RTT from OLT1 to ONU$_{2i+1}$, $W_j^{[2i-1]}$ and $W_j^{[2i]}$ are the jth window size for ONU$_{2i-1}$ and ONU$_{2i}$ respectively, and R_D is the transmission speed.

5. PERFORMANCE EVALUATIONS

In this section, the system performances of the proposed multi-OLT PON with the FS bandwidth allocation scheme and the LS bandwidth allocation scheme are analysed and compared with the existing single-OLT PON in terms of successive Grant scheduling time, cycle time evolution, and average packet delay for uniform and non-uniform traffic loads. Finally, we have investigated the throughput of the multi-OLT PON and compared with the single-OLT PON for non-uniform traffic load and LS bandwidth allocation scheme. For the variation of traffic model (considering maximum packet size) between FTTH terminals and CHs of WSN, performances of the multi-OLT PON system are also analysed using different packet sizes for FTTH terminals and WSN by considering different maximum granted window sizes, $W^{[max]}$.

The random packet-based simulation model was used considering tree topology based PON architecture with two OLTs and 16 ONUs, where eight ONUs (ONU1, ONU3 ... ONU15) were considered for OLT1 and other eight ONUs (ONU2, ONU4 ... ONU16) were for OLT2. The distances from OLTs to ONUs were assumed as random to range from 10 to 20 km [15]. The downstream transmission and upstream transmission speeds of both were 1 Gb/s. Highly bursty random traffic patterns were generated for non-uniform traffic condition. For uniform traffic condition every ONU was considered to generate an offered load while for non-uniform traffic condition every ONU was considered to generate any value from 0 to the offered load. Our simulations took into account queuing delay, transmission delay, propagation delay, and processing delay. The simulation scenario is summarized in Table 1.

Table 1. Simulation scenario.

Symbol	Explanation	Value
N_{ONU}	Total Number of ONUs	16
N_{FTTH}	Number of ONUs for FTTH terminals	8
N_{WSN}	Number of ONUs for WSN	8
N_{OLT}	Number of OLTs	2
D	Distance between OLTs and ONUs (random)	10-20 km
T_{max}	Maximum cycle time	2 ms
T_g	Guard time for single OLT PON	5 μs
R_D	Transmission speed	1Gb/s
T_{proc}	Processing time	10μs
B	Packet size	1500 byte (FTTH) 1024 byte (WSN)
B_{eth}	Ethernet overhead	304 bits
B_{rep}	Report message size	576 bits
P_{max}	Maximum transmission window	20 packets
$W^{[max]}$	Maximum granted window	$(B_{eth} + B)*P_{max}$

Firstly, the impact of modified version of interleaved polling algorithm on the proposed multi-OLT PON is investigated. In the simulation, random packet sizes but not larger than $W^{[max]}$ are generated to every ONU for the LS scheme, however, it is fixed to $W^{[max]}$ for every ONU with the FS scheme even if the network load is low. For simplicity, we assume that the starting time of the 1st Grant message by OLT1 to ONU1 is at 0.0 μs. Figure 6 shows the comparison of successive Grant scheduling time between the single-OLT PON and the multi-OLT PON for both of the FS and the LS bandwidth allocation algorithms. From these results, it is clear that successive Grant scheduling time of the multi-OLT PON with interleaved polling algorithm provides about 80μs less time delay at the 16th ONU for both of the LS scheme and the FS scheme, because the proposed multi-OLT PON does not require any guard time between every two successive ONUs.

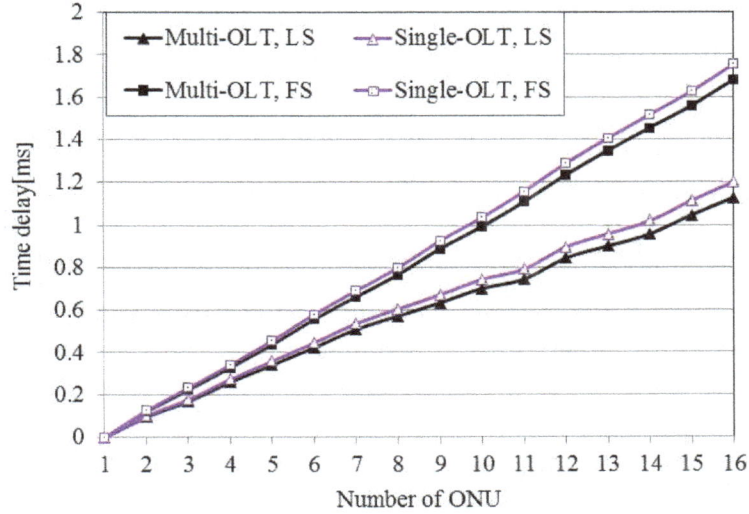

Figure 6. Comparison of successive Grant scheduling time between the single OLT PON and the multi-OLT PON systems

Cycle time is an important parameter to calculate packet delay in PON system. Usually, data packets of all ONUs of present cycle are sent to the OLT in the next cycle, because all ONUs still have to inform the OLT for their queue status during the present cycle. Depending on the traffic generation time in an ONU, transmission starting delay will be on average half of a cycle time, T_{cycle}.

For the FS bandwidth allocation scheme, cycle time T_{cycle} will be constant for all traffic loads [16]. In this case, data will suffer from the same delay in every cycle and it does not depend on the present network traffic. From the Table 1, it is clear that the cycle time for the FS scheme is always 2.0 ms. If the traffic load is very low then bandwidth utilization problem [17] will be severe for the FS scheme due to the bandwidth wastage, Ethernet overhead, and guard time in the single-OLT PON. However, the proposed multi-OLT PON can provide better bandwidth utilization under the same traffic condition due to the avoidance of guard time. The following formula represents the cycle time for the multi-OLT PON with the FS scheme.

$$T_{cycle} = \frac{N.B_{rep} + N.(B + B_{eth}).P_{max}}{R_D} \tag{7}$$

where, B_{rep}, B, and B_{eth} denote the Report message size, packet size, and Ethernet overhead respectively, N represents the number of ONUs, P_{max} indicates the maximum transmission window, and R_D is the transmission speed.

In the LS bandwidth allocation scheme, cycle time is variable under low and non-uniform traffic condition, since the granted window size is based on the requested window size. Bandwidth utilization in the LS scheme will be improved than the FS scheme for both of the single-OLT and multi-OLT PONs due to avoidance of bandwidth wastage. Equation (8) represents the cycle time for the multi-OLT PON with the LS scheme.

$$T_{cycle} = \frac{N.B_{rep} + \sum_{i=1}^{N}(B+B_{eth}).P_i}{R_D} \tag{8}$$

where, P_i is the granted transmission window for ONU_i and $P_i \leq P_{max}$.

To calculate the average packet delay in the LS scheme, consideration of the bursty nature of cycle time is important, because it is clear that the maximum queuing time of aggregated packets in an ONU depends on the cycle time of the previous cycle. Figure 7 gives an idea about the bursty nature of the cycle time for the LS scheme under non-uniform traffic condition for both of the single-OLT and the multi-OLT PONs. However, the cycle time for FS bandwidth allocation scheme is constant. The figure also clearly shows the existence of longer and shorter cycle times. Since the cycle times of previous cycles influence the data packet delay of present cycle, consideration of this cycle time variation influence the end-to-end packet delay and data queuing time in ONUs.

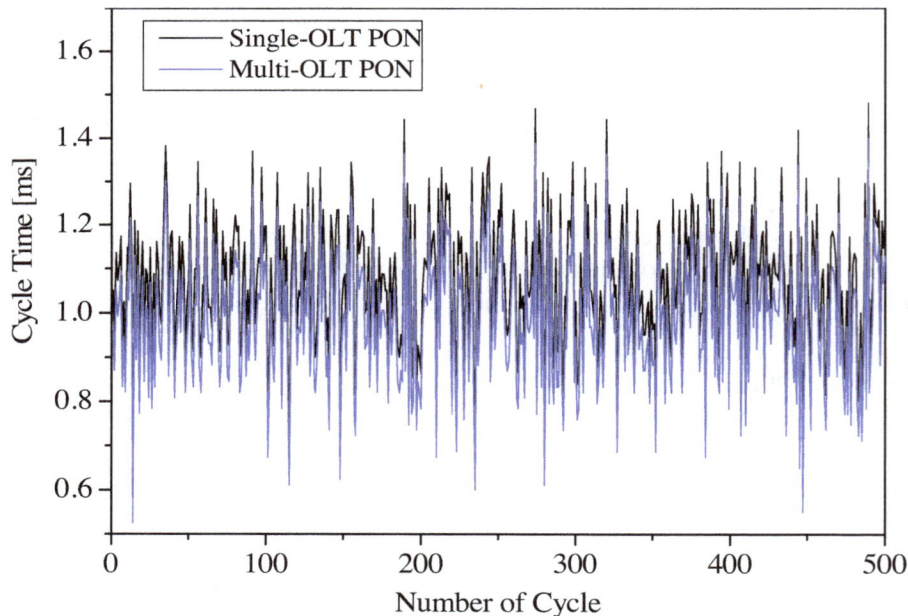

Figure 7. Evolution of cycle times for the LS scheme

Figure 8 compares the average end-to-end packet delay vs. offered load among the single-OLT PON and the proposed multi-OLT PON under uniform traffic conditions. In this simulation, same packet length is considered for both of FTTH terminals and CHs of WSN. The results show that delay characteristics are similar among the single-OLT PON and the multi-OLT PON for both of the FS scheme and the LS scheme. Small amounts of delay improvement about <0.1ms at low traffic condition and <0.2ms at high traffic condition are observed for the multi-

OLT PON due to the avoidance of guard time. Therefore, the multi-OLT PON can support the existing FS scheme and LS scheme with better delay efficiency than the single-OLT PON.

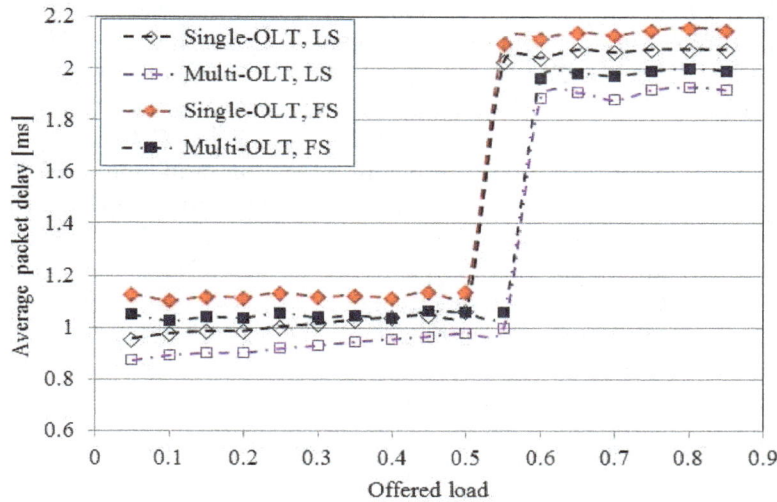

Figure 8. Comparison of average packet delay for uniform traffic load

Figure 9 compares the average end-to-end packet delay vs. offered load between the single-OLT PON and the multi-OLT PON for the LS scheme under non-uniform traffic conditions. For the single-OLT PON, the average packet delay increases sharply when the offered load changes from 0.5 to 0.55, whereas for the multi-OLT PON, such a sharp increase of the average packet delay is observed at the offered load from 0.6 to 0.65. The reason is similar with the case of uniform traffic load in Figure 8, and the proposed multi-OLT PON provides better delay efficiency than the single-OLT PON even under non-uniform traffic conditions.

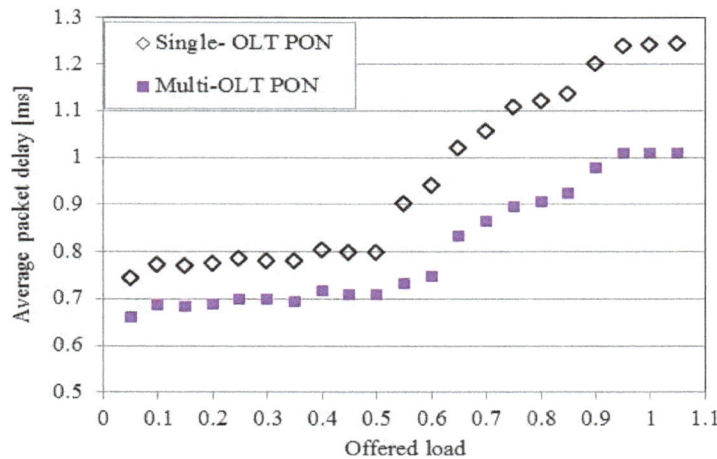

Figure 9. Comparison of average packet delay for non-uniform traffic load with the LS scheme

From the analysis of end-to-end packet delays for both of uniform and no-uniform traffic conditions it is clear that for higher traffic loads the average packet delay increases very sharply for both of the single-OLT and multi-OLT PONs. Because the aggregated traffic load becomes the determining factor and end-to-end packet delay increases quickly due to congestion. Due to the congestion, if a packet cannot be sent in its first requested transmission window, it will have an influence over two or multiple cycle times.

By now, considering same packet length for FTTH terminals and CHs of WSN, the multi-OLT PON presented here achieves better delay efficiency than the single-OLT PON. The maximum packet length for FTTH terminals and WSN are 1500 bytes [16] and 1024 bytes [18] respectively. When same bandwidth (50% bandwidth ratio) is allocated for both networks, then the networks will suffer from different average packet delays at the offered load ≥0.55 to ≤0.75 (difference of offered load is 0.2) as shown in Figure 10. Here, Figure 10 shows the average end-to-end packet delay vs. offered load characteristics under uniform traffic conditions using different bandwidth allocation ratios among FTTH terminals and CHs of WSN. To reduce this difference, we should choice the proper bandwidth allocation ratio among FTTH terminals and CHs of WSN. Similarly, if 70% and 30% bandwidth are used for FTTH terminals and CHs of WSN respectively then the difference of offered load becomes 0.3 (≥0.45 to ≤0.75). On the other hand, if 60% and 40% bandwidth ratio is used then the difference of offered load reduced to 0.05 (≥0.6 to ≤0.65). Considering these cases, 60% and 40% bandwidth for FTTH terminals and CHs of WSN respectively provide optimized average packet delay for both networks. This bandwidth ratio reflects the ratio of packet length of FTTH terminals and WSN if the packet length is changed then the bandwidth ratio will also be changed.

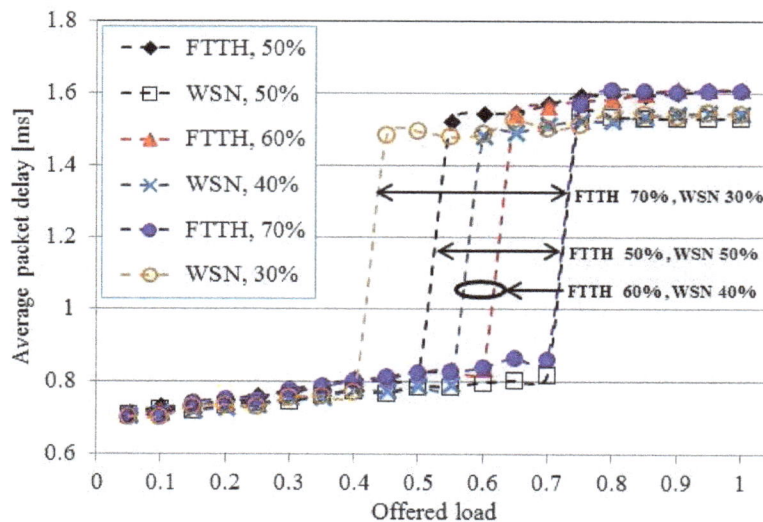

Figure 10. Average packet delay for uniform traffic load with the LS scheme

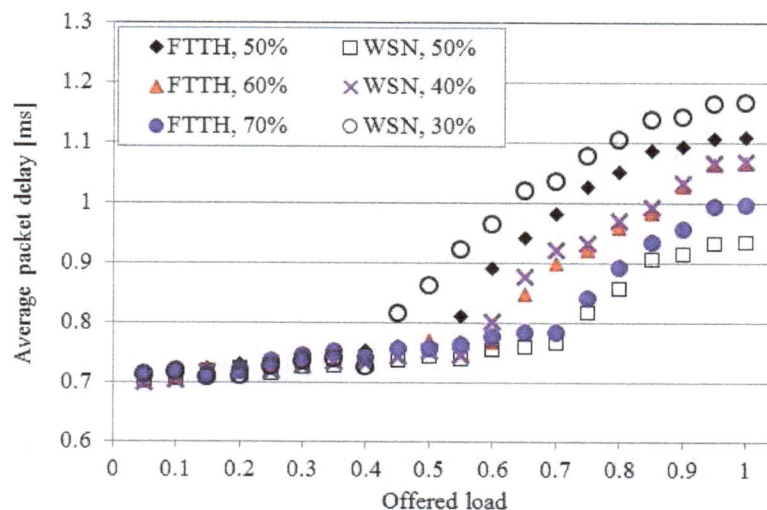

Figure 11. Average packet delay for non-uniform traffic load with the LS scheme

Figure 11 shows the average end-to-end packet delay vs. offered load characteristics under non-uniform traffic conditions using different bandwidth allocation ratios. From the results, it is clear that the optimized average packet delay for non-uniform traffic is also achieved at 60% and 40% bandwidth for FTTH terminals and CHs of WSN respectively.

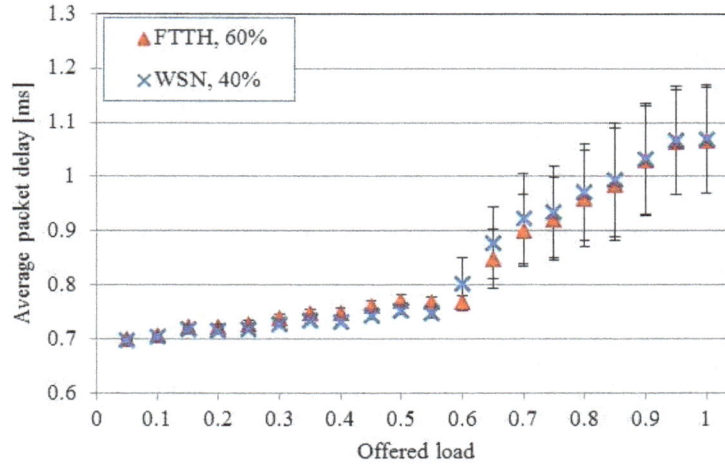

Figure 12. Analysis of error bars for non-uniform traffic load with the LS scheme

We provide the error bars analysis using standard deviation for non-uniform traffic condition for optimized (60% bandwidth for FTTH terminals and 40% bandwidth for CHs of WSN) bandwidth allocation ratio among FTTH terminals and CHs of WSN of the multi-OLT PON. Figure 12 shows the average packet delay vs. offered load characteristic with error bars using standard deviation for non-uniform traffic condition with the LS bandwidth allocation scheme. From the analysis of error bars it can be mentioned that with the increase of offered load standard deviation is also increased. However, the maximum deviation is not more than 20% at maximum offered load of 1.0.

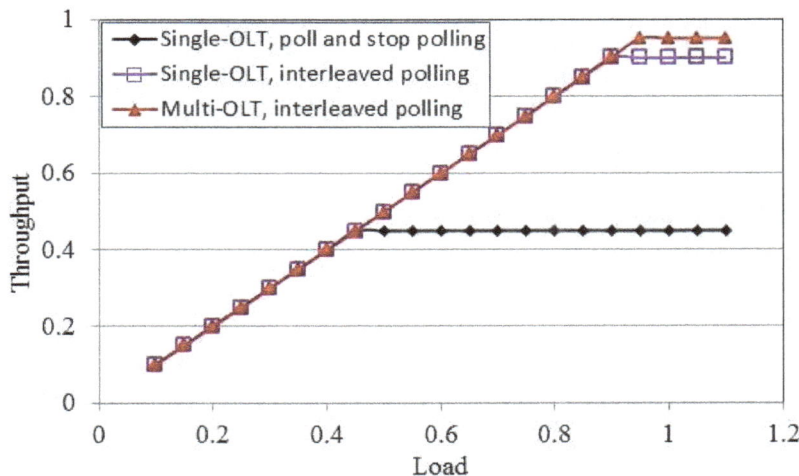

Figure 13. Comparison of throughput

Finally, we provide the comparison of throughput among the single-OLT polling algorithms and the multi-OLT interleaved polling algorithm. Figure 13 shows the throughput of the single-OLT poll and stop polling algorithm, the single-OLT interleaved polling algorithm, and the multi-OLT interleaved polling algorithm. As shown in the figure, the single-OLT poll and stop polling algorithm has the lowest throughput (less than 50%) due to the wastage of bandwidth as idle time, whereas the single-OLT and the multi-OLT interleaved polling algorithms overcome this bandwidth wastage problem to improve the channel utilization. Moreover, the multi-OLT interleaved polling algorithm achieves a throughput of 95% (compared with the single-OLT interleaved polling algorithm having 90%), which is due to the bandwidth savings by avoiding guard time between every two successive ONUs.

6. CONCLUSIONS

In this paper, we proposed a multi-OLT PON structure for both FTTH terminals and WSN. A modified interleaved polling algorithm and scheduling algorithm of a control message were also proposed. From the computer simulation results, evaluating average packet delay, it is found that the proposed structure can reduces the end-to-end latency about 0.1ms up to the offered load of 0.6 while 0.2ms at offered load of 1.0 for both uniform and non-uniform traffic loads. Moreover, the multi-OLT PON can accommodate 5% more traffic load than the single-OLT PON without suffering from congestion. Therefore, it can be mentioned that the multi-OLT PON can effectively connects CHs of large WSN to make a converged network with FTTH terminals of a u-City with latency efficiency. The analysis performed in this study using the existing FS and LS bandwidth allocation schemes for both uniform and non-uniform traffic conditions proves the validity of the proposed multi-OLT PON structure. Moreover, the proposed multi-OLT PON structure with modified interleaved polling algorithm also outperforms other existing single-OLT polling algorithms in term of throughput for non-uniform traffic load with LS scheme. Furthermore, it is found that the convergence of FTTH and WSN in a multi-OLT PON is an efficient approach in that it provides cost effective solution than using two separate PONs, less packet delay, improved bandwidth utilization, and throughput under both uniform and non-uniform traffic conditions.

ACKNOWLEDGEMENTS

This work was supported in part by the JSPS-NRF bilateral joint research project.

REFERENCES

[1]   Shinya Ito and Kenji Yoshigoe, (2009) "Performance Evaluation of Consumed-Energy-Type-Aware Routing (CETAR) for Wireless Sensor Networks" *International Journal of Wireless & Mobile Networks (IJWMN)*, vol. 1, no. 2, pp 90-101.

[2]   Shio Kumar Singh, M P Singh, and D K Singh, (2010) "Energy Efficient Homogenous Cluster - ing Algorithm for Wireless Sensor Networks" *International Journal of Wireless & Mobile Networks(IJWMN)*, vol.2, no.3, pp 49-61.

[3]   Jun Tang, Xinyu Jin, Yu Zhang, Xianmin Zhang, Wenyu Cai, (2007) "A Hybrid Radio Over Fiber Wireless Sensor Network Architecture," *Proc. of International Conference on Wireless Communications, Networking and Mobile Computing,WiCom2007*, pp 2675-2678.

[4]   Monir Hossen, Byung-Jun Jang, Ki-Doo Kim, and Youngil Park, (2009) "Extension of Wireless Sensor Network by Employing RoF based 4G Networks," *Proc. of the 11th International Conference on Advanced Communication Technology(ICACT)*, pp 275-278 .

[5]   Monir Hossen, Ki-Doo Kim, and Youngil Park, (2010) "Efficient Clustering Algorithm for Delay Sensitive PON-based Wireless Sensor Network," *Proc. of International Conference on Optical Internet,* pp 341-343.

[6]   G. Kramer, B. Mukherjee, and G. Pessavento, (2002) "IPACT: A Dynamic Protocol for an Ethernet PON (EPON)," *IEEE Communication Magazine*, vol.40, issue 2, pp 74-80.

[7]   Monir Hossen, Ki-Doo Kim, and Youngil Park, (2010) "Synchronized Latency Secured MAC Protocol for PON Based Large Sensor Network," *Proc. of the 11th International Conference on Advanced Communication Technology(ICACT)*, pp 1528-1532.

[8]   Masaki Tanaka, Takashi Nishitani, Hiroaki Mukai, Seiji Kozaki, and Hideaki Yamanaka, (2011) "Adaptive Dynamic Bandwidth Allocation Scheme for Multiple-Service in 10G-EPON System," *Proc. of IEEE ICC 2011*.

[9]   G. Kramer and B. Mukherjee, (2001) "Ethernet PON: Design and Analysis of an Optical Access Network," *Photonic Network Communication*, Vol. 3, no. 3, pp 307-319.

[10]   G. Kramer, B. Mukherjee, S. Dixit, Y. Ye, R. Hirth, (2002) "Supporting Differentiated Classes of Service in Ethernet Passive Optical Networks," *Journal of Optical Networks*, vol.1, no.8, pp 280-298.

[11]   M. Ma, Y. Zhu, and T. H. Cheng, (2003) "A Bandwidth Guaranteed Polling MAC Protocol for Ethernet Passive Optical Networks," *Proc. of IEEE INFOCOM*, San Francisco, CA, pp 22-31.

[12]   Chadi M. Assi, Yinghua Ye, Sudhir Dixit, Mohamed A. Ali, (2003) "Dynamic Bandwidth Allocation for Quality-of-Service Over Ethernet PONs," *IEEE Journal on Selected Areas in Communications*, vol. 21, no.9, pp 1467-1477.

[13]   IEEE Draft P802.3ah™/D1.2, (2002) "Media Access Control Parameters, Physical Layers and Management Parameters for Subscriber Access Networks," http://grouper.ieee.org/groups/802/3/efm.

[14]   Bjorn Skubic, Jiajia Chen, Jawwad Ahmed, Lena Wosinska, and Biswanath Mukherjee, (2009) " A Comparison of Dynamic Bandwidth Allocation for EPON, GPON, and Next-Generation TDM PON" *IEEE Communication Magazine*, pp S40-S48.

[15]   I-Shyan Hwang, Zen-Der Shyu, Liang-Yu Ke, Chun-Che Chang, (2008) " A Novel Early DBA Mechanism with Prediction-based Fair Excessive Bandwidth Allocation Scheme in EPON," *Journal of Computer Communications*, vol. 31, no. 9, pp 1814–1823.

[16]   B. Lannoo, L. Verslegers, D. Colle, M. Pickavet, P. Demeester, and M. Gagnaire, (2007) "Thorough Analysis of the IPACT Dynamic Bandwidth Allocation Algorithm for EPONs," *Proc. of Fourth International Conference on Broadband Communications, Networks and Systems*, pp 486-494.

[17]   Kyuho Son, Hyungkeun Ryu, Song Chong, (2004) "Dynamic Bandwidth Allocation Schemes to Improve Utilization Under Non-uniform Traffic in Ethernet Passive Optical Networks," *Proc. of IEEE International Conference on Communications*, vol. 3, pp 1766–1770.

[18]   V. Rajendran, K. Obraczka, and J.J. Garcia-Luna-Aceves, (2006) " Energy-Efficient, Collision-Free Medium Access Control for Wireless Sensor Networks," *Journal of Wireless Networks*, vol. 12, no. 1, pp 63–78.

14

PERFORMANCE ANALYSIS OF WLAN STANDARDS FOR VIDEO CONFERENCING APPLICATIONS

Lachhman Das Dhomeja[1], Shazia Abbasi[1], Asad Ali Shaikh[1], Yasir Arfat Malkani[2]

[1]Institute of Information and Communication Technology, University of Sindh, Jamshoro, Pakistan
[2]Institute of Mathematics and Computer Science, University of Sindh, Jamshoro, Pakistan
lachhman@usindh.edu.pk, shazia.abassi@usindh.edu.pk, asad.shaikh@usindh.edu.pk and,yasir.malkani@usindh.edu.pk

ABSTRACT

A number of traffic characterization studies have been carried out on wireless LANs, which indicate that the wireless settings pose major challenges, especially for high bandwidth and delay sensitive applications. This paper aims to evaluate a number of Quality of Service (QoS) parameters related to video conferencing over three major WLAN Standards 802.11a, 802.11b and 802.11g. To study the traffic characterization behaviour of these WLAN standards, we have simulated the environment for each of these standards and performed experiments. Results are verified through the delivery of successful H.261 video traffic import in OPNET-14 Network simulator. We found that a trade-off exists between the selected data rate, physical characteristics and the frequency spectrum (number of channels) for every standard. The traffic of video conferencing is characterized over each standard in terms of delay performance, traffic performance and load and throughput performance. The results show that quality of video traffic is a function of the frequency band, physical characteristic, maximum data rate and buffer sizes of WLAN standards.

KEYWORDS

WLAN standards, QoS support, video conferencing, OPNET-14.

1. INTRODUCTION

The field of wireless local area networks (WLANs) is being widely studied and used in various emerging research domains such as mobile and pervasive computing, where WLANs provide high-speed wireless connection and support accessing information from anywhere and anytime. WLANs[3-8] support a wide range of applications, which may include simple applications such as web browsing, file transferring, etc and the other ones, for instance, real-time multimedia applications (e.g., video streaming and video conferencing). The latter requires better quality of service than the former. A detailed survey of quality of service in WLANs can be found at [1]. While simple applications may well be supported by WLANs, the applications requiring better quality of services (real-time multimedia applications) may suffer due to reasons that the wireless channels are error prone, band-limited, etc [2].

IEEE 802.11[3-8] is a vital standard for wireless LAN, which adopts the standard 802 logical link control (LLC) protocol that is further divided into two sub layers: physical layer (PHY) and medium access control (MAC) layer. This configuration provides optimized functionality for wireless communication. Initially 802.11 had two physical layers: Direct Sequence Spread Spectrum (DSSS) and Frequency Hopping Spread Spectrum (FHSS) and later on the physical layer was categorized into three types with different physical characteristics and frequency spectrum [3].

The physical characteristic of 802.11a [5] and 802.11g [8] are identical – both are based on OFDM and support data rate of 54 Mb/s. However they differ in operating frequency spectrum – 802.11a operates on 5 GHz band, while 802.11g on 2.4 GHz. 802.11b [4] is based on DSSS and operates at 2.4 GHz band with transmission rate from 1 to 11 Mb/s. 802.11a has significant advantage due to the wide range spectrum of 5 GHz, having more number of independent channels. Both 802.11b and 802.11g are compatible with each other as both operates on 2.4 GHz spectrum, but this may cause degradation in system performance as 2.4 GHz is a small band spectrum with a lesser number of independent channels.

 The main objective of the work presented in this paper is to study the performance of three WLAN standards, 802.11a, 802.11b and 802.11g, especially when supporting a video conferencing application, using these parameters: (i) used frequency spectrum and available number of orthogonal channels for each WLAN standards, (ii) used modulation technique of each standard, (iii) selected buffer size for application, (iv) Load of control and data channels in each standard. We have used OPNET-14 simulator [9-10] to simulate 802.11a/b/g-based WLANs for our study.

The remainder of this paper is organized as follows. Section 2 provides related work. Section 3 describes experimental setup in which we discuss how WLAN has been setup. Section 4 presents and discusses the results of various performance tests we have conducted. Finally, section 5 concludes the paper.

2. RELATED WORK

There exists a large body of research on Multimedia Traffic characterization either on wired or wireless LANs, such as [11,12,13,14,15,16,17,18,19,20]. In [11], video traffic has been analysed on Ethernet LANS over two different data rates: 10 Mbps and 100 Mbps focusing on characterization of quality of video in terms of glitches. The research efforts [12,13,14,15,16] focus on 802.11b network, where in [12], authors have characterized UDP traffic over 802.11b WLANs using parameters such as throughput, average delay, frame error rate, IP loss rate, etc. In contrast, in [13], the 802.11b has been investigated for its capabilities for voice traffic with the focus on minimizing Mean Opinion Score (MOS) requirements.

The authors [14] have developed a simple packet delay jitter analytical model for IEEE 802.11 DCF, which computes average packet delay and packet delay variability. The authors in [15] have extended their work carried out in [14] in which the proposed model is used to evaluate the performance of WLANs, especially for applications involving both voice and data. The parameters being used for performance evaluation include throughput, jitter, and loss rate probability.

In [16], an analytical model has been developed for IEEE 802.11b Distributed Coordination Function (DCF), which calculates various parameters such as an average voice packet, voice packet delay variation (jitter) and packet drop probability for voice packets. Additionally, authors have studied the impact of data transmission on voice capacity. Work carried out in [17] focuses on addressing the issues of real-time video streaming over WLANs, especially over

IEEE 802.11b. Their solution is based on combination of forward error control (FEC) coding with the ARQ protocol.

The authors [18] have investigated IEEE 802.11e standard for its capability for QoS support. This is done by evaluating both the Enhanced Distributed Channel Access (EDCA) and the Polling-based Channel Access modes of this standard for multiple traffics such as real-time audio and video traffic. Similarly, [19] also focuses on evaluation of WLAN standard's capability for QoS support and involves evaluation of two MAC layer protocols: DCF (Distributed Coordination Function) and EDCF (Enhanced Distributed Coordination Function). Their evaluation suggests that EDCF is better in providing QoS for multiple services environment as EDCF has a capability to distinguish and prioritize services. The authors [20] have evaluated the performance of 802.11 WLAN in terms of throughput, using four types of applications, http, remote login, video conferencing and voice over IP. Evaluation of throughput is done in presence of high priority traffic (video conferencing, voice over IP traffic) and low priority traffic (http, remote login traffic).

It can be noted that research efforts discussed above provide performance evaluation of a single WLAN standard. In contrast to these, our study provides performance analysis of three WLAN standards: 802.11a /g /b for video conferencing application.

3. EXPERIMENTAL SETUP

In order to study the performance of three Wireless LAN standards for video conferencing application, we have simulated the network setup using OPNET-14 simulator and conducted various tests on it. A basic infrastructure mode network has been used for experimental setup, in which four Basic Service Sets (BSSs): BBS 0, BSS 1, BSS 2 and BSS 3 have been set, where every BSS is working as independent wireless LAN. Multiple number of wireless clients are running under BSS 0 and BSS 1, a wireless application server is running on BSS 2 and BSS 3 is configured as a backbone network for connecting other three LANs.

These three LANs, BSS 0, BSS 1 and BSS 2 are connected to each other with three routers. Each router has two WLAN interfaces; one of them serves as an access point for BSS 0, BSS 1 and BSS 2, while the other interface of three routers make up the WLAN-backbone (BSS 3). The first interface, IF0 of BSS 0, BSS 1 and BSS 2 is configured as an access point with BSS ID being set to 0, 1 and 2 respectively. Whereas the second interface, IF1 of three BSSs (i.e., BSS 0, BSS 1 and BSS 2) have been disabled for access point functionality and all of IF1s have been set with the same ID, which is 3. These three IF1s make up a Wireless backbone (BSS 3), as mentioned before.

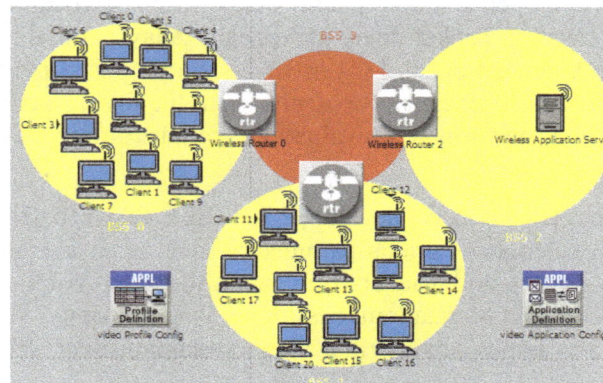

Figure 1: WLAN Setup

Attribute Setup: The attributes of each standard adopted according to the requirements. The most common parameters we used are tabulated below (table 1)

Table 1: Important parameters and their setting for the basic simulation

Attributes	Settings
WLAN Standards	802.11a / 802.11b / 802.11g
Data Rate	54Mbps / 11 Mbps / 54 Mbps
Physical characteristic	OFDM / Direct Sequence/ OFDM
Frequency spectrum	5 GHz / 2.4 GHz / 2.4 GHz
Transmit Power	0.005 W
Packet Reception Power Threshold	-95
Buffer size	256000
Max. Receiver Life time	0.5 Sec
Roaming capability	Disabled
Beacon Interval	0.02 Sec
Short time limit	7
Long time limit	4

4. RESULTS AND DISCUSSION

In this section, we present the results of various tests we have conducted to analyze the performance of three wireless standards, 802.11 a, 802.11 b and 802.11g. Tests include Delay Performance, Traffic Performance and Load and Throughput Performance.

4.1 Delay Performance

Delay is an essential metric to characterize the QoS of any network, especially for real time Multimedia application. The delay is defined as the time taken by the system for data to reach the destination after it leaves the source. The delay for any network can be measured at three layers, end-to-end delay, wireless LAN delay and MAC (media access control) delay. Wireless LAN delay depends on used frequency band and media access delay on media access technique and physical characteristic of the standard, while end-to-end delay includes both wireless LAN delay and MAC delay. The following figures show the results of end-to-end delay test, wireless delay test and MAC delay test.

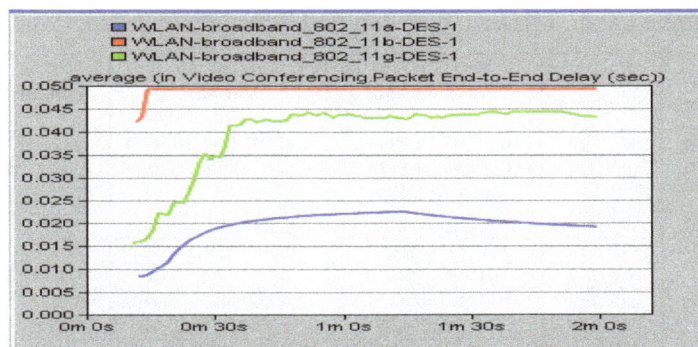

Figure 2: End-to-end delay of three standards

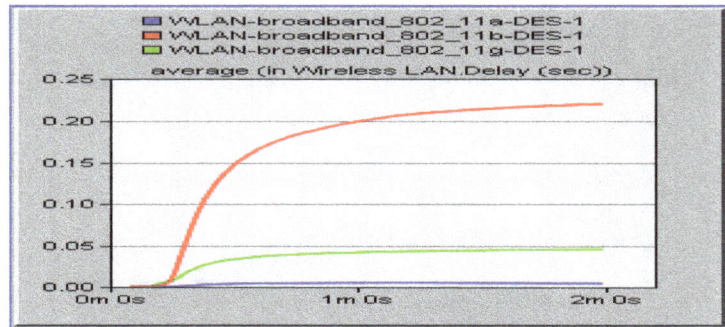

Figure 3: Wireless LAN delay of three standards

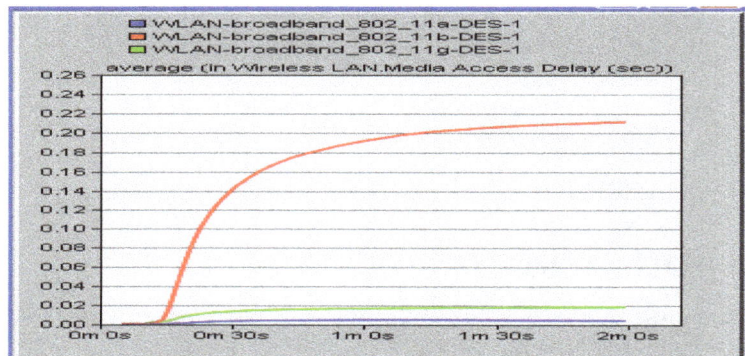

Figure 4: Media access delay of three standards

Results presented in figure 2, figure 3 and figure 4 indicate that the 802.11a has minimum delay, while the 802.11g has twice as much delay as in 802.11a, whereas 802.11b has maximum delay in all three cases. While in all three cases 802.11b has maximum delay, in end-to-end delay (figure 2) the performance of 802.11b is improved a little in comparison with other two cases (WLAN delay and MAC delay) because DSSS works efficiently with minimum orthogonal channels of 2.4 GHz. To summarize, all three results suggest that 80211a performs better than other two standards. We have also calculated sample mean, variance and standard deviation of all three tests (end-to-end delay, wireless LAN delay and MAC delay) for each standard, and the results are summarized in table 2.

Table 2. Sample mean, variance and standard deviation of performance delay test

Delay Performance	WLAN Standard	Sample mean	Variance	Standard deviation
Packet end to end Delay	802.11a	0.0196348008932	1.12652781103E-005	0.00335637871973
	802.11b	0.0491222211525	8.98037442955E-007	0.000947648375166
	802.11g	0.0397200175033	6.31130731542E-005	0.00794437367916
Wireless LAN delay	802.11a	0.00480128812775	2.87035722714E-006	0.00169421286359
	802.11b	0.169570800291	0.0046528157452	0.0682115514059
	802.11g	0.037609206252	0.000175778509105	0.0132581487812
Media Access Delay	802.11a	0.00439140768535	2.44990949064E-006	0.00156521867183
	802.11b	0.16274496359	0.0043023502159	0.0655923030233
	802.11g	0.0151322222921	2.70551934843E-005	0.00520146070679

4.2 Traffic Performance

One of the parameters that can influence on overall performance of the Wireless Local Area Networks (WLANs) is traffic analysis. Traffic analysis includes traffic sent, traffic dropped and traffic received. Traffic sent determines the capability of the system to transmit amount of data from the source point, while traffic received determines the amount of the data received at the destination. The traffic drop in applications such as video conferencing is often caused by the buffer overflow and the amount of data dropped can be determined from the amount of data transmitted and received. We have conducted various tests for traffic performance of three wireless standards and the following figures (Figures 5, 6 and 7) show the results of these tests.

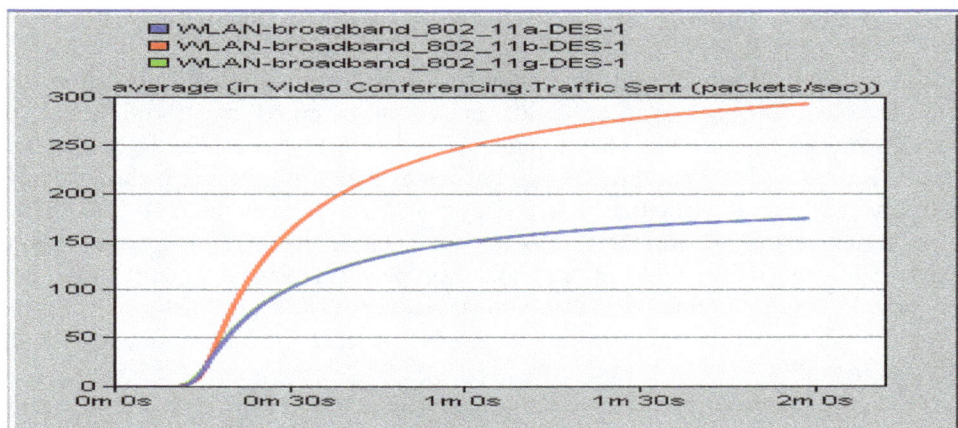

Figure 5: Video conferencing: traffic sent (packets/sec) of three standards

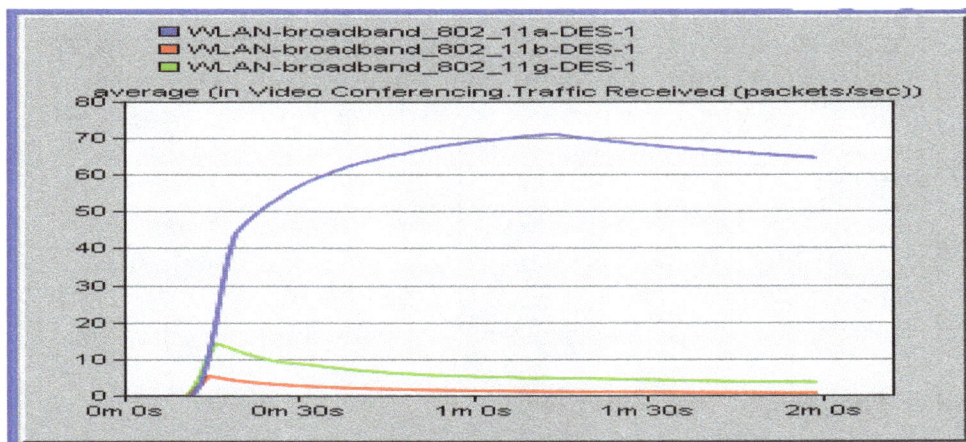

Figure 6: Video conferencing: traffic received (packets/sec) of three standards

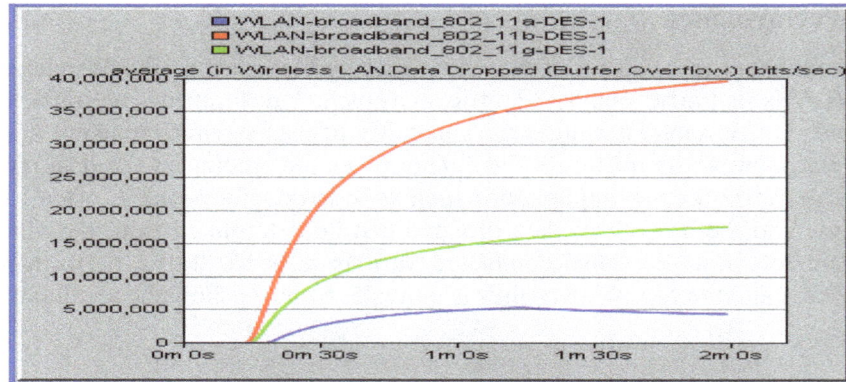

Figure 7: Video conferencing: data dropped (bits/sec) of three standards

To study the traffic performance of three wireless standards, the same amount of data was inputted to the system for each of three standards. As can be noted from the figures 5, 6 and 7, the 80211a has minimum data drop as compared to both 802.11g and 802.11b, hence maximum receipt. As compared to 802.11b, the 802.11g has lesser data drop. Figure 6 shows that 802.11a receives around 65% more data than the 802.11g and 70 % than 802.11b. With regard to capability of transmission of data, the results (figure 5) show that 802.11b performs better than 802.11a and 802.11g, while 802.11a and 802.11g perform almost equally. We have also calculated sample mean, variance and standard deviation of all three traffic performance tests (traffic sent, traffic received and traffic dropped) for each standard, and the results are summarized in the table 3.

Table 3. Sample mean, variance and standard deviation of traffic performance test

Traffic Performance	WLAN Standard	Sample mean	Variance	Standard deviation
Traffic Sent (packets /sec)	802.11a	0.0196348008932	1.12652781103E-005	0.00335637871973
	802.11b	0.0491222211525	8.98037442955E-007	0.000947648375166
	802.11g	0.0397200175033	6.31130731542E-005	0.00794437367916
Traffic Received (packets /sec)	802.11a	0.0048012881277	2.87035722714E-006	0.00169421286359
	802.11b	0.169570800291	0.0046528157452	0.0682115514059
	802.11g	0.037609206252	0.000175778509105	0.0132581487812
Traffic dropped (buffer over flow (bits/sec)	802.11a	0.0043914076853	2.44990949064E-006	0.00156521867183
	802.11b	0.16274496359	0.0043023502159	0.0655923030233
	802.11g	0.0151322222921	2.70551934843E-005	0.00520146070679

4.3 Load & Throughput Performance

Another parameter that influences the overall performance of the wireless standards is load & throughput. The load & throughput test is concerned with the receipt of the payload data without considering overhead of network against load. We have conducted three tests to analyse the load & throughput performance of each of three wireless standards: load carried by the system (figure 8), throughput of the system for offered load (figure 9) and retransmission attempts until either packet is successfully transmitted or it is discarded as a result of reaching short or long retry limit (figure 10).

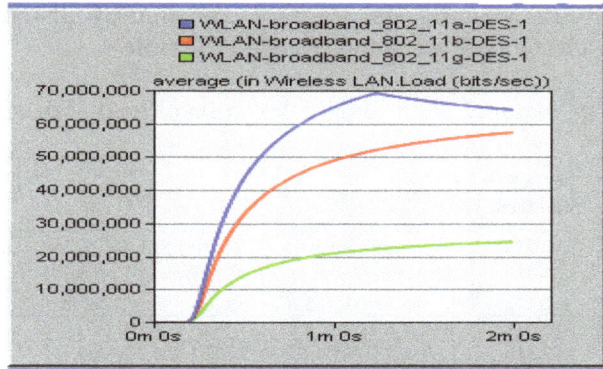

Figure 8: Wireless LAN load (bits/sec) of three standards

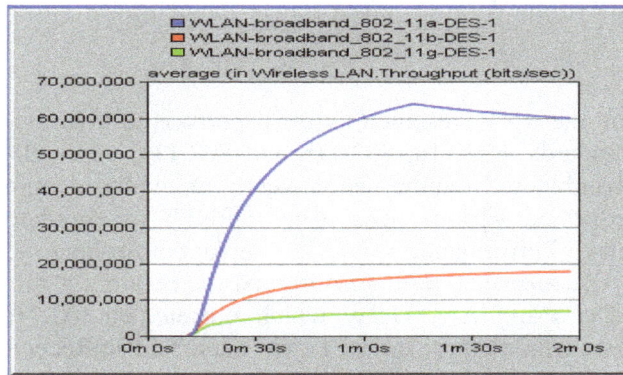

Figure 9: Wireless LAN Throughput (bits/Sec) of three standards

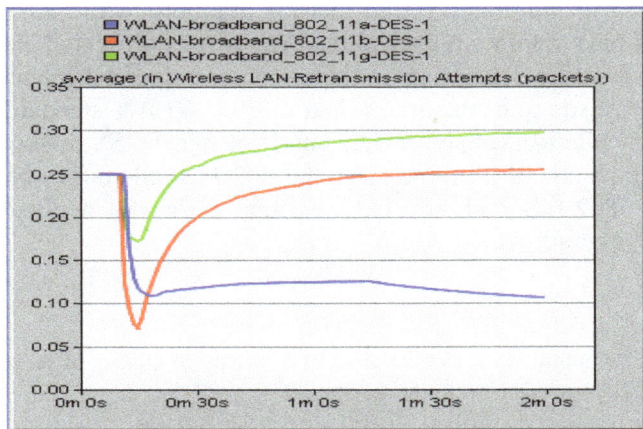

Figure 10: Wireless LAN Retransmission Attempts (Packets) of three standards

The purpose of this test is to identify which standard is an efficient under heavy load of system in terms of higher throughput with minim number of retransmission attempts. These plots show that the performance of 802.11a in a heavily loaded network is better than both 802.11b and

802.11g. The sample mean, variance and standard deviation of all three load & throughput performance tests for each standard are summarized in the table 4.

Table 3. Sample mean, variance and standard deviation of load & throughput performance

Load & Throughput Performance	WLAN Standard	sample mean	variance	standard deviation
Average Load (bits/sec)	802.11a	50,657,352.2283695	553,011,813,641,622	23,516,203.2148394
	802.11b	40,047,690.898171	367,763,492,205,956	19,177,160.6919783
	802.11g	17,185,538.0215591	65,634,393,372,658.8	8,101,505.62381208
Average Throughput (bits /sec)	802.11a	46,958,600.1995828	472,677,785,964,715	21,741,154.2003803
	802.11b	12,936,731.3084012	32,561,925,057,069.1	5,706,305.72762002
	802.11g	5,183,099.62016764	4,606,459,734,051.55	2,146,266.46389761
Average Retransmission attempts (Packs)	802.11a	0.127212416114	0.00106241441837	0.0325946992373
	802.11b	0.225022059352	0.00194173949542	0.0440651732712
	802.11g	0.273108083123	0.000952249981415	0.0308585479473

5. CONCLUSION

Main motivation behind the work presented in this paper was to investigate the performance of three main WLAN standards, 802.11a, 802.11b and 802.11g, especially for the applications which have high bandwidth requirements such as video conferencing application. Consequently, we performed various tests using OPNET-14 simulator. Performance tests conducted were Delay Performance, Traffic Performance and Load & Throughput Performance. In Delay Performance test, we observed the results for three cases: End-To-End Delay, Wireless LAN Delay and MAC Delay, which indicate that 802.11a has minimum delay. Traffic performance test included three cases: Traffic sent, Traffic Received and Data Dropped. The results of this test showed that the 80211a has minimum data drop, hence improved data receipt. Load & Throughput test includes three cases: WLAN load, Throughput and Retransmission Attempts. We observed that under heavy load of LAN traffic, 802.11a has maximum throughput with minimum retransmission attempts, while 802.11g performs poorly under traffic load and have minimum throughput. The results presented clearly indicate that the performance of WLAN varies depending on the selection of parameters such as used frequency band, physical characteristic and maximum data rate of WLAN standards. We observed that OFDM is an efficient while working on 5 GHz band whereas DSSS performs better on 2.4 GHz band. Considering the results of all three tests, the 802.11a falls out to be a better choice than two other standards, 802.11b and 802.11g, especially for the applications requiring high bandwidth for smooth operations.

REFERENCES

[1]   H. Zhu, M. Li, I. Chlamtac, B. Prabhakaran, " A survey of quality of service in IEEE 802.11 networks", IEEE Wireless Communications, 2004.

[2]   M. Chen and A. Zakhor, "Rate Control for Streaming Video over Wireless", IEEE Conference Proceedings, INFOCOM, 2004.

[3]   IEEE 802.11 WG, Part 11: Wireless LAN Medium Access Control (MAC) and Physical Layer (PHY) specification, Standard, IEEE, August 1999.

[4]   IEEE 802.11b WG, Part 11: Wireless LAN Medium Access Control (MAC) and Physical Layer (PHY) specification: High-speed Physical Layer Extension in the 2.4 GHz Band, IEEE, September 1999.

[5]   IEEE 802.11a WG, Part 11: Wireless LAN Medium Access Control (MAC) and Physical Layer (PHY) specification: High-speed Physical Layer in the 5GHz Band, September 1999.

[6]   IEEE 802.11e WG, Draft Supplement to Part 11: Wireless Medium Access Control (MAC) and physical layer (PHY) specifications: Medium Access Control (MAC) Enhancements for Quality of Service (QoS), IEEE Standard 802.11e/D3.3.2 , November 2002.

[7]   IEEE 802.11e/D11.0, Draft Supplement to Part 11: Wireless Medium Access Control (MAC) and physical layer (PHY) specifications: Medium Access Control (MAC) Enhancements for Quality of Service (QoS), October 2004.

[8]   IEEE Standard 802.11g/D1.1-2001, Part11: Wireless LAN Medium Access Control (MAC) and Physical Layer (PHY) specifications: Further Higher-Speed Physical Layer Extension in the 2.4 GHz Band.

[9]   OPNET Technologies, http://www.opnet.com/solutions/network_rd/modeler.html

[10]   Mohammad M. Siddique, A. Konsgen, "WLAN Lab Opnet Tutorial", University Bermen Press, 2007.

[11]   F. A. Tobagi and I. Dalgic, "Performance evaluation of 10base-t and 100base-t Ethernets carrying multimedia traffic", IEEE Journal on Selected Areas in Communications, 14(7), 1996, pp. 1436–1454.

[12]   M. Arranz, R. Aguero, L. Munoz, and P. Mahonen, "Behavior of udp-based applications over IEEE 802.11 wireless networks" , 12th IEEE International Symposium on Personal, Indoor and Mobile Radio Communications, 2001, pp. 72–77.

[13]   D. P. Hole and F. A. Tobagi, "Capacity of an IEEE 802.11b WLAN supporting VoIP", IEEE Proceedings of ICC-4, 2004.

[14]   P. Raptis, V. Vitsas, P. Chatzimisios and K. Paparrizos," Delay jitter analysis of 802.11 DCF", Electronics Letters, Issue Date: Dec. 6 2007.

[15]   P. Raptis, V. Vitsas, P. Chatzimisios and K. Paparrizos , "Voice and Data Traffic Analysis in IEEE 802.11 DCF Infrastructure WLANs", Advances in Mesh Networks, 2009. MESH 2009. Second International Conference, page(s): 37 - 42, June 2009.

[16]   C. Brouzioutis, V. Vitsas and P. Chatzimisios, "Studying the Impact of Data Traffic on Voice Capacity in IEEE 802.11 WLANs", Communications (ICC), 2010 IEEE International Conference
Issue Date: 23-27 May 2010.

[17]   A. Majumda, D.G. Sachs, I.V. Kozintsev, K. Ramchandran, M. M. Yeung, "Multicast and Unicast Real-time Video Streaming over Wireless LANs," IEEE Trans. Circuits Sys. Video Tech., vol. 12, June 2002, pp. 524-534.

[18]   D. Chen, D. Gu and J. Zhang, "Supporting Real-time Traffic with QoS in IEEE802.11e Based Home Networks", IEEE Proceedings of Consumer Communications and Networking Conference (CCNC-04), 2004.

[19]   J. Sengupta, G. Singh Grewal, "Performance Evaluation of IEEE 802.11 MAC layer in supporting Delay sensitive Services", International journal of wireless & mobile networks (IJWMN), Vol: 2, No.1, Feb 2010.

[20]   K. Sharma, N. Bhatia, N. Kapoor, "Performance Evaluation of 802.11 WLAN Scenarios in OPNET Modeler" , International Journal of Computer Applications (0975 – 8887) Volume 22– No.9, May 2011.

15

PERFORMANCE ANALYSIS AND IMPLEMENTATION FOR NONBINARY QUASI-CYCLIC LDPC DECODER ARCHITECTURE

Tony Tsang[1]

[1]Department of Computer Engineering, La Trobe University, Melbourne, Australia.

ABSTRACT

Non-binary low-density parity check (NB-LDPC) codes are an extension of binary LDPC codes with significantly better performance. Although various kinds of low-complexity iterative decoding algorithms have been proposed, there is a big challenge for VLSI implementation of NBLDPC decoders due to its high complexity and long latency. In this brief, highly efficient check node processing scheme, which the processing delay greatly reduced, including Min-Max decoding algorithm and check node unit are proposed. Compare with previous works, less than 52% could be reduced for the latency of check node unit. In addition, the efficiency of the presented techniques is design to demonstrate for the (620, 310) NB-QC-LDPC decoder.

KEYWORDS

Wireless Communications; NB-QC-LDPC; VHDL; Performance Analysis

1. INTRODUCTION

Channel coding plays key role in providing a reliable communication method that can overcome signal degradation in practical channels. A new field of study into non-algebraic codes based on linear transformations using generator and parity check matrices are led off the breakthrough of convolutional codes [1]. Using a finite-state process is encoded the Convolutional codes, which generates them a linear order encoding scheme. Afterward, convolutional codes led to the discovery of a class of codes called Turbo codes [2], which are the class of concatenated convolutional codes and randomize the order of some of the bits by using an inter-leaver. The first known capacity approaching error correction code is Turbo code, which provides a powerful error correction capability when decoded by an iterative decoding algorithm. The rediscovery of low density parity check (LDPC) code, which was originally proposed by Gallager [3] and was later generalized as MacKay-Neal [4] code puts back Turbo coding as the forward error correction (FEC) technique. LDPC codes were neglected for a long time since their computational complexity for the hardware technology was high. LDPC codes have acquired considerable attention due to its near capacity error execution and powerful channel coding technique with an adequately long code-word length. There are several advantages LDPC codes over Turbo codes. In the decoding of Turbo codes it is difficult to apply parallelism due to the sequential nature of the decoding algorithm, while in LDPC decoding can be accomplished with a high degree of parallelism to attain a very high decoding throughput. Since, turbo codes usually cause a large delay, but LDPC codes do not need a long inter-leaver. LDPC codes can be constructed directly for a desired code rate. In case of turbo codes, which are based on convolutional codes, require other methods such as puncturing to acquire the desired rate.

The codes are classified into two major categories, explicitly, block codes and convolutional codes. Hamming codes, Bose-Chaudhuri-Hocquenghem (BCH) codes, Reed-Solomon (RS) codes and newly rediscovered LDPC codes are the example of block codes. Block codes like Hamming, BCH and RS codes have structures but with limited code length. A bounded-distance decoding algorithm is usually employed in decoding block codes except LDPC codes, in general it is hard to use soft decision decoding for block codes.

Advances in error correcting codes have revealed that, using the message passing decoding algorithm, irregular LDPC codes can accomplish consistent communication at SNR very close to the Shannon limit on the AWGN channel, outperforming turbo codes of the same block size and code rate. Hou et al [5] examined the numerical analysis method for calculating the threshold of the LDPC codes. For the AWGN channel; the proposed method is implemented to the uncorrelated flat Rayleigh fading channel. Additionally, using the nonlinear optimization technique of differential evolution, the degree distribution pairs are optimized for the uncorrelated Rayleigh fading channel and observe that their threshold values are very close to the capacity of this channel for moderate block size with excellent performance. Zhang et al [6] proposed the two adaptive coded modulation schemes employing LDPC codes for Rayleigh fading channels. The proposed schemes have made good use of the time-varying nature of Rayleigh fading channel. It is also observed that the proposed schemes perform better by employing LDPC with large code length.

The performance of irregular LDPC codes is investigated in [7] with three BP based decoding algorithms, specifically the uniformly most powerful (UMP) BP- based algorithm, the normalized BP-based algorithm, and the offset BP-based algorithm on a fast Rayleigh fading channel by employing density evolution. It is observed from the study of proposed method that the performance and decoding complexity of irregular LDPC codes with the offset BP-based algorithm can be very close to that with the BP algorithm on the fast Rayleigh fading channel. Ohhashi and Ohtsuki [7] provide successful evolution of irregular LDPC codes, and then analyze the performance of regular LDPC codes with the normalized BP-based algorithms on the fast Rayleigh fading channel. Formulas for short and long regular LDPC codes are derived based on the probability density function (PDF) of the initial likelihood information and DE for the normalized BP-based algorithm on the fast Rayleigh fading channel. Performance of the long regular LDPC codes with the normalized BP-based algorithm in the proposed method outperforms the BP algorithm and the UMP BP-based algorithm on fast Rayleigh fading channel. In this paper new Nonbinary LDPC codes have been developed and implement the newly designed codes on FPGA platform. Simulation results demonstrate that the proposed Nonbinary LDPC codes achieve a 0.8 dB coding gain over randomly constructed codes and perform 1 dB from the Shannon-limit at a BER of 10−6 with a code rate of 0.89 for block length of 620.

2. NONBINARY QC-LDPC CODES AND MIN-MAX DECODING ALGORITHM

2.1. Non-binary QC-LDPC Codes

StandardLet GF (q) be a finite field with q elements. A q -ary LDPC code C is given by the null space of a sparse parity-check matrix H = [hi;j] over GF(q) . If each column has constant weight (the number of nonzero entries in a column) and each row has constant weight γ (the number of nonzero entries in a row), the code C is referred to as a (γ, ρ) -regular LDPC code. If the columns and/or rows of the parity-check matrix have multiple weights, the null space of H gives an irregular LDPC code.

Consider the construction field GF (q) with α as a primitive element. Then $\alpha^{-\infty}=0, \alpha^0=1, \alpha, \ldots \alpha^{q-2}$

give all the elements of GF(q) and $\alpha^{q-1}=1$. We define α-multiplied circulate permutation matrix (CPM) as a $(q-1)\times(q-1)$ matrix over GF (q) . Each row of the α-multiplied CPM is the right cyclic-shift of the row above it multiplied by α ; the first row is the right cyclic-shift of the last row multiplied by α . Thus, each α-multiplied CPM is characterized by its offset, which denotes the position of the nonzero entry in the first row of the matrix. Generally, for $0 \leq l < q-1$, the nonzero entry in the l th column of the α-multiplied CPM is α^l . For example, an α-multiplied CPM over GF(8) with an offset of 3 is shown as

$$\begin{bmatrix} 0 & 0 & 0 & \alpha^3 & 0 & 0 & 0 \\ 0 & 0 & 0 & 0 & \alpha^4 & 0 & 0 \\ 0 & 0 & 0 & 0 & 0 & \alpha^5 & 0 \\ 0 & 0 & 0 & 0 & 0 & 0 & \alpha^6 \\ \alpha^0 & 0 & 0 & 0 & 0 & 0 & 0 \\ 0 & \alpha & 0 & 0 & 0 & 0 & 0 \\ 0 & 0 & \alpha^2 & 0 & 0 & 0 & 0 \end{bmatrix}$$

If **H** is an array of α-multiplied CPMs and all-zero matrices, then the null space of gives a non-binary quasi-cyclic (QC)- LDPC codes. The non-binary QC-LDPC codes are usually constructed based on algebraic methods [8]. QC-LDPC codes are advantageous over other codes in terms of encoding complexity.

2.2. B. Min-Max Decoding Algorithm

A bipartite graph called Tanner graph, represented graphically the non-binary LDPC code C, which consists of two disjoint sets of nodes. Nodes in one set, called variable nodes (VNs), represent the code symbols; nodes in the other set, called check nodes (CNs), represent the check-sums that the code symbols must satisfy. For a code with a J × n parity-check matrix, label the VNs from 0 to n − 1 and the CNs from 0 to J − 1 . The i th CN is connected to the j th VN by an edge if and only if hi;j $\leq$ = 0 . The VNs connected to the i th CN simply correspond to the code symbols that are contained in the i th check-sum. The number of these VNs is referred to as the CN degree of thei th CN. The CNs connected to the j th VN simply correspond to the check sums that contain the j th VN. The VN degree of the j th VN referred to the number of these CNs.

Message passing algorithms can iteratively decode the Non-binary LDPC codes. Instead of a single message (as for the binary codes), a vector of sub-messages are passed through each edge of the Tanner graph. The QSPA is approximated the min-max decoding algorithm [9] with reduced complexity and 0.1 V 0.2 dB of performance degradation, and thus is widely adopted for decoder implementation [10, 11].

For $0 \leq i < J$ and $0 \leq j < n$, we define $N_i = \{j : 0 \leq j < n, h_{i,j} \neq 0\}$, and $J_j = \{i : 0 \leq i < J, h_{i,j} \neq 0\}$. Let Kmax be the maximum number of iterations to be performed. Let Lj be the a priori information of the j th code symbol from channel. Let $\mathbf{z} = (z_0, z_1, \ldots, z_{n-1})$ be the hard decision symbols for code symbols generated either when initialized or during the VN processing. For $0 \leq j < n$, each $\mathbf{L}_j = (L_{j,0}, L_{j,1}, \ldots, L_{\alpha^{q-2}})$ consists q of log likelihood ratios (LLRs) $\mathbf{L}_{j,\alpha^l} = log(Prob(z_j = \beta)) / log(Prob(z_j = \alpha^l))$

where $0 \leq l < q - 2$, or $l = -\infty$, and is the most likely symbol for zj (i.e., $Prob(z_j = \beta)$ is the largest among all q probabilities). From this definition, all LLRs are non-negative. Also, the smaller the $\mathbf{L}_{j,\alpha^l}$ is, the most likely that the code symbol zj is α^l. Let $\mathbf{Q}_j = (Q_{j,0}, Q_{j,1}, \ldots, Q_{j,\alpha^{q-2}})$ be the a posteriori information of the j th code symbol. Let $\mathbf{Q}_{j \to i} = (Q_{j \to i,0}, Q_{j \to i,1}, \ldots, Q_{j \to,\alpha^{q-2}})$ and $\mathbf{R}_{j \leftarrow i} = (R_{j \leftarrow i,0}, R_{j \leftarrow i,1}, \ldots, R_{j \leftarrow,\alpha^{q-2}})$ be the VN-to-CN and CN-to-VN message vectors passed between the j th VN and i th CN, respectively. The VN-to-CN and CN-to-VN messages are also referred to as extrinsic messages. Let be the sequence of finite field element assignments of $z_{j'}(j' \in N_i \setminus j, z_{j'} \in GF(q))$ such that

$$\sum_{j' \in N_i \setminus j} h_{i,j'} z'_j = h_{i,j} \alpha^l$$

Let be the iteration counter. The min-max decoding is as follows.

Initialization : Set k = 0 . For all $0 \leq j < n$, set Qj = Lj , and zj = arg max_l (Qj ;_l) . For all i ; j , set Qj→i = Lj .

Step 1) Parity check: Compute the syndrome $\mathbf{zH^T}$ of z. If $\mathbf{zH^T} = 0$, stop decoding and output as the decoded code word; otherwise go to Step 2.

Step 2) If k = Kmax , stop decoding and declare a decoding failure; otherwise, go to Step 3.

Step 3) CN processing: Compute the CN-to-VN message $R_{j \leftarrow i,\alpha^l} = min_{z_j \in L_i, z_j = \alpha^l} (max_{j' \in N_i \setminus j} Q_{j' \to i, z_{j'}})$ and pass messages from CNs to VNs.

Step 4) VN processing: $k \leftarrow k + 1$. Compute the VN-to-CN message in two steps. First compute the primitive message by

$$Q_{j \to i,\alpha^l} = L_{j.\alpha^L} + \sum_{i' \in Jj \setminus i} R_{j \leftarrow i',\alpha^l} .$$

Then we compute

$$Q^{min}_{j \to i} = min_{\alpha^l \in GF(q)} Q_{j \to i,\alpha^l}$$

After that, the VN-to-CN message is normalized

$$Q_{j \to i,\alpha^l} = Q_{j \to i,\alpha^l} - Q^{min}_{j \to i}$$

and passed from VN to CN. Also, we update the reliability of each code symbol by

$$Q_{j \to i,\alpha^l} = L_{j.\alpha^L} + \sum_{i' \in Jj} R_{j \leftarrow i,\alpha^l} .$$

The code symbol is determined as $max_{\alpha^l}(Q_{j,\alpha^l})$. Go to Step 1.

3. ARCHITECTURAL IMPLEMENTATION OF ALGORITHM

In this brief, the proposed scheme is designed a check node unit (CNU), which perform consists of two major parts: a sorter that sorts out the 1.5nm v-to-c messages with the smallest nonzero LLRs, and a path constructor that generates the c-to-v messages from the sorting results. In the following, the architectures for these two parts are presented.

3.1. Sorter Architecture

The right-shift cells [12] can realized with the sorter. Totally processing elements (PEs) are required. Each PE composes of a right-shift cell and a compare cell. The right-shift cell one is responsible for the date storage and right-shift operation. The right-shift cell executes the comparison and generates the control signals for the right-shift cell. Here, ri denotes the right-shift enable signal, pi presents the pre-sorted data, and is the result of comparison. Finally, the sorter will output nm intrinsic messages with the most significant magnitudes.

As a result, the decode processing algorithm of CNU is implemented, and two sub-blocks are needed. In total, the compare sub-block consists of | L | (dc −1) –input comparators and one (| L | −1) -input comparator, where | L | is the cardinality of the set L(c | av = _i). The inputs data can be read off from the LUT of the finite solution sequence generator, which simple control unit achieved with selecting. The output to the sorter provides the final result, which will pick up the nm most significant ones, realign them in decreasing order, and output them to the v-th VNU for further operations. The corresponding hardware structure of the CNU block is presented in Fig. 2(a) as above. Moreover, the proposed architecture can be further simplified as shown in Fig. 2(b), where only one (dc − 1) -input comparator is employed. For both structures, given   all | L | finite solution sequences are generated by the module illustrated in Fig. 1. In Fig. 2(a), firstly all possible solutions are processed in parallel to select the required one. In the outer loop, indexed with i, only nm intrinsic messages with the most significant magnitudes are chosen by data sorter which is shown in Fig. 1. In Fig. 2(b), the switch is left open during the first step. All solution sequences are input in serial. A 2-input comparator with one delay element is employed to generate LLR value of $R_{cv}^{k,t}(\alpha^i)$. In the outer loop, the switch is closed and the sorter will pick up the nm most significant intrinsic messages.

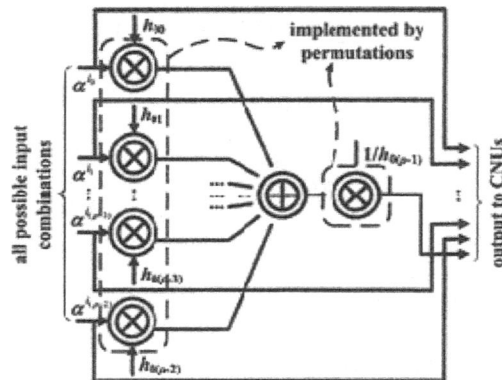

Fig. 1. Internal structure of generator for possible solution sequences

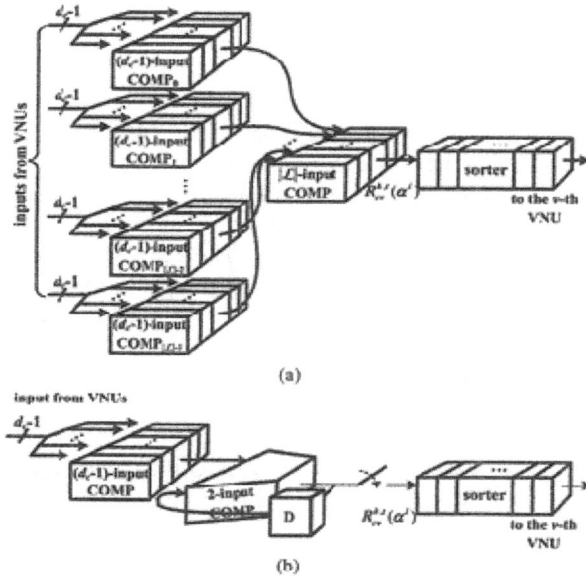

Fig. 2. Propoased CNU block architecture employing data sorter

The Sorter is defined in VHDL in the following:
entity Sorter is
port (dc − 1; input : in std logic vector;
input : in std logic vector(31downto0);
ri; pi; ci; nm : out std logic vector(31downto0);
architecture eq1 of Sorter is
component Right − Shifter the data of
inputs to detect the largest value • • • ;
− − Smallest values is right shift with the same
number of difference between the two input;
component Swap − −Swap the two input
from initial state to finalstate;);

• • • ;

component Compare the data of
inputs to detect the largest value • • • ;

− − largest values is sorter with the output $R_{cv}^{k,t}(\alpha^l)$

3.2. Path Constructor

The architecture of the path construction module is illustrated in Fig. 3. The 1.5nm sorted v-to-c messages are read from the RAM device one at a cycle, starting from the smallest nonzero LLR. During the path construction for c-to-v message Rm;n; e(i) , which is the index of the variable node that the message belongs to, is passed to the decoder to generate a binary vector with dc bits, in which only the e(i) th bit is "1" . The bit test block in Fig. 4 takes this vector and Pj . It outputs "1" when Pk (e(i)) ≤ = 1 , as the enable signal to other blocks. If it outputs "0" , the path has been constructed before and thus should not be included. The f in the proposed algorithm can be computed by two GF adders. The multiplexer is added to enable the algorithm initialization. If the enable signal is "1", a new path will be constructed. In addition, the new path vector and the finite symbol will be stored into a path track block and a symbol track block, respectively. In addition, the message LLR, the computed path symbol, and the path signal will be sent to select units (SU), which is used to compute Rm;n for each variable node. In an SU, the path symbol is added to the z (n) , which is the GF symbol of the zero-LLR node in column n . The computed symbol fn is copied to the GF comparator and the first-in-first-out (FIFO) buffer consisting of serially concatenated registers. In this way, each symbol in FIFO can be simultaneously compared with the newly computed fn to test if fn ∈ fLn . The GF comparator outputs "1" when fn equals none of the symbols in the FIFO. When the GF comparator and the path signal both output "1" ,h the LLR and GF symbol fn will be loaded into corresponding memory devices, and a new entry of Rm;n is computed. Otherwise, the message is skipped. The path construction will take 4 nm cycles in total, according to the proposed algorithm. Using the proposed CNU architecture, only 1.5 nm sorted v-to-c messages need to be stored for each check node. Compared with sorting dc nm intermediate messages for each of the check node process using the FB scheme, the memory requirement has been substantially reduced. Once the sorted messages are available, the path constructor can start to derive the c-to-v messages for the current check node. The total cycles needed to finish the check node processing are around (2dc + 5:5nm). The architecture can compute all dc c-to-v messages at a time. The total cycles for a check node process are much smaller compared with the original path construction architecture in [13].

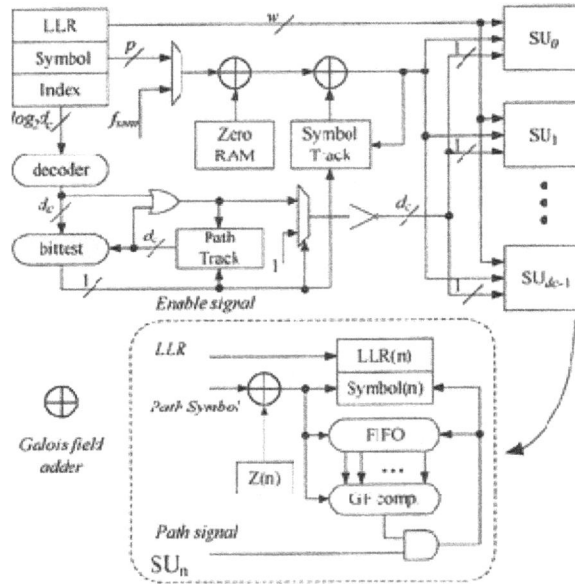

Fig. 3. Architecture of the path constructor

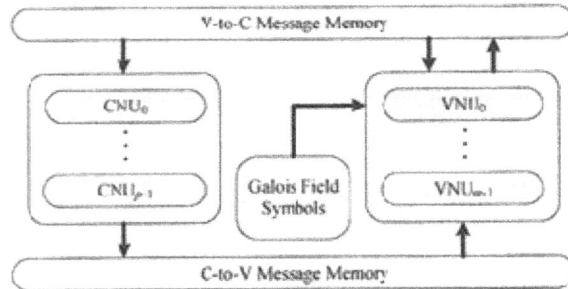

Fig. 4. Top-level architecture of the min-max NB-LDPC decoder

The path constructor is defined in VHDL in the following:

```
entity Path Constructor is
port Rm:n; e(i); V to CMessage : in std logic vector;(31 down to 0);
Clk : in std logic;
dc; C to V Message : out std logic vector(31downto0);
end Path Constructor;
architecture Decoder of Path Constructor is
signal e(i) : std logic vector(31downto0);
component Decoder the Decocder in Path Constructor
port e(i) : in std logic vector(31downto0);
port dc : out std logic vector(31downto0);
end component;
component GF adder the Decoder in Path Constructor • • • ;
component Multiplexer
– – signal initializerfor real multiplier
port P; fsum : instd logic vector(31downto0);
Clk : in std logic;
port (Zero RAM; Symbol Track : instd logic vector(31downto0);
data; dc : out std logic vector (31downto0);
end component;

• • • ;

component Select Unit the SU in Path Constructor
port LLR; Path Symbol; Path Signal : in std logic vector;
(31 down to 0);
port fn : out std logic vector (31downto0);
end component;
component GF Comparator the SU in Path Constructor • • • ;
component GF adder the SU in Path Constructor • • • ;
component FIFO the SU in Path Constructor • • • ;
end component;

• • • ;
```

3.3. Min-Max Decoder Architecture

Section In our design, layered decoding is adopted to reduce the memory requirement and to increase the decoding convergence speed. The H matrix is divided into several layers. The c-to-v messages derived from one layer are used right away to update the v-to-c messages of the next layer.

The CNU mainly consists of the proposed sorter and path constructor. The variable node unit (VNU) is an extension of that used in binary LDPC decoders and can directly be implemented with an adder, a subtractor, and a parallel sorter. Since only nm messages are kept for each vector, it is possible that, for a message in one vector, there is no message with the same GF element in the other vector. Taking this into account, the variable node elementary processing is mainly composed of two loops of nm cycles each to skim through all the values of the two input vectors. The details of the processing and VNU architecture can be referred to [14].

The top-level architecture of the min V max decoder is shown in Fig. 4. Here, we assume that there are p CNUs and m VNUs in the proposed decoder architecture.

The parameter p is determined by the row number of one layer, and m is equal to the number of columns with nonzero GF elements in one layer. For QCNB-LDPC codes, the H matrix can be divided into several submatrices of dimension s×s , and each column of H has at most one nonzero entry in each layer. Hence, the parameters are p = s; m = s×dc. During the check node processing, all the rows in one layer are processed in parallel. During the decoding iterations, the p CNUs will read channel messages from the v-to-c message memory and fill the c-to-v message memory with updated c-to-v messages. The VNUs will compute updated v-to-c messages once check node processing has been done. Denote the v-to-c LLRs of layer l in the j th decoding iteration by $Q_l^{(j)}$. Represent the c-to-v LLRs computed from layer l in the j th iteration by $R_l^{(j)}$. It can be derived that $Q_{l+1}^{(j)} = (Q_l^{(j)} + R_l^{(j)}) - R_{l+1}^{(j-1)}$.

After v-to-c messages have been updated to the v-to-c message memory, a new round of check node processing will begin. In this brief, based on the proposed architecture, a decoder for a (620, 310), (6, 3) NB-LDPC code over GF(32) is designed. The base matrix size is 31×31 . The check node degree and variable node degree are 6 and 3, respectively. There are 31 CNUs and 186 VNUs as well as two message memories.

The Min-Max Decoder is defined in VHDL in the following:

entity Min − Max Decoder is
port (VtoC Message : in std logic vector(31downto0);
CtoV Message : out std logic vector(31downto0););
end Min − Max Decoder;
architecture Decoder of Min − Max Decoder is
signal Data : std logic vector(31downto0);
signal nm : std logic vector(31downto0);
signal pi : std logic vector(31downto0);
signal mi : std logic vector(31downto0);
component H matix − −Parameterizable • • • ;
component GF Symbol − −Parameterizable • • • ;
component CNU − −Sorter & Path constructor • • • ;
component VNU − −Adder & Subtractor & Parallel sorter • • • ;

3. PERFORMANCE SIMULATION RESULT

The study of how changes in performance depend on changes in parameter mode values is known as sensitivity analysis. We can vary some parameter's value a little, and see its influence degree to the model's performance, for example, the throughput or response time. Throughput is an action-related metric showing the rate at which an action is performed at steady-state. In other words, the throughput represents the average number of the activities completed by the system during one unit time.

From Figure 5, it can be observed that the impact of the number of devices on the throughput of transmit is more sensitive than the FB-Min-Max, nm = 32, floating and nm = 16 , floating modulation for nm = 16 . If we could make some efforts to optimize the cache, and raise the Min-Max, nm = 32 , floating modulation form 0.6 to 0.85 or even more high value, the throughput of transmit could greatly improved. In Fig. 6, some simulation results of the frame error rate of the minVmax algorithm for an (620, 310) NB-LDPC code over GF(32) using layered decoding are shown. Here, we use w = 5 bits to represent each LLR, and nm = 16 most reliable entries are kept in each message. The maximum number of iterations is set to 15. It can be observed that under 5-bit quantization, the proposed trellis path scheme has the same performance as the FB scheme and only has about 0.02-dB performance loss compared with the floating point minVmax decoding. For the purpose of comparison, the FB minVmax algorithm with nm = 32 and the fast fourier transform (FFT)-BP algorithm are also included.

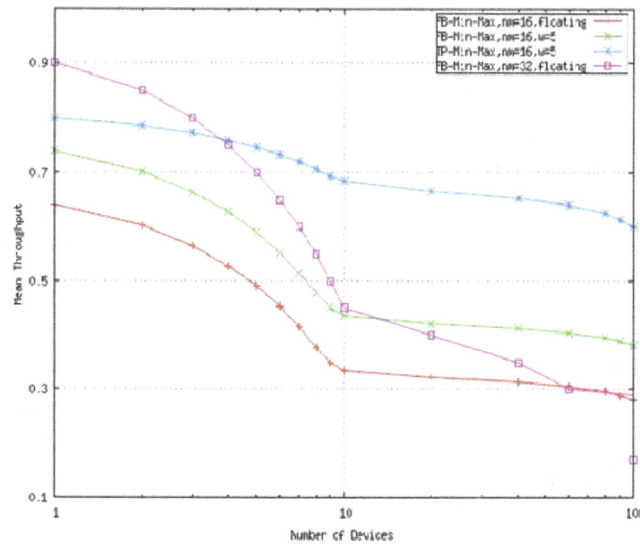

Fig. 5. Throughput versus Number of Devices

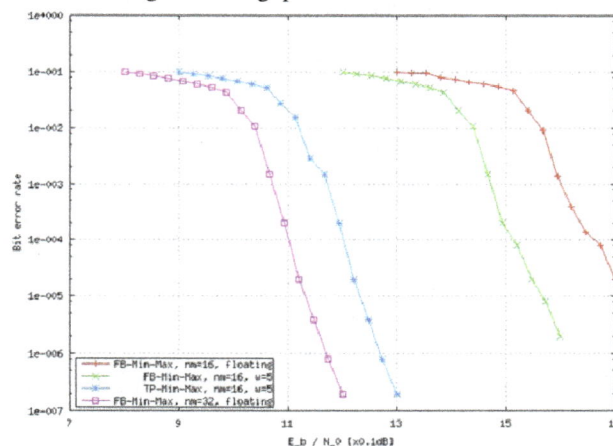

Fig. 6. Performance simulation for a (620,310) NB-LDPC code

4. CONCLUSIONS

Channel coding plays key role in providing a reliable communication method that can overcome signal degradation in practical channels. In this paper new Non-binary LDPC codes have been developed and implement the newly designed codes on FPGA platform. Prototype architecture of the LDPC codes has been implemented by writing Hardware Description Language (VHDL) code and targeted to VLSI chip. Simulation results demonstrate that the proposed Non-binary QC-LDPC codes achieve a 0.8 dB coding gain over randomly constructed codes and perform 1 dB from the Shannon-limit at a BER of 10−6 with a code rate of 0.89 for block length of 620.

REFERENCES

[1]   Charles H. L., Error-Control Convolutional Coding, 1st edition, Artech House, Inc. Norwood, MA, USA, 1997.

[2]   Berrou C., Glavieux A. and Thitimajshima P., "Near Shannon limit error correcting coding and decoding: turbo-codes", IEEE ICC'93, Geneva, Switzerland, pp. 1064-1070, 1993.

[3]   Gallager R. G., Low-Density Parity-Check Code. Cambridge, MA: MIT Press, 1963.

[4]   Mackay D. and Neal R., "Near Shannon Limit Performance of Low Density Parity Check Codes", Electronic Letters, Vol. 32, No.18, Pages 1645-1646, 1996.

[5]   Huo G. and Alouini S., "Another Look at the BER Performance of FFH/BFSK with Product Combining Over Partial-Band Jammed Rayleigh-Fading Channels", IEEE Transaction on Vehicular Technology, vol. 50(5),Pages 1203-1215, 2001.

[6]   Zhang H. and Moura J. M., "The design of structured regular lpdc codes with large girth", GLOBECOM, Pages 4022V4027, 2003.

[7]   Ohhashi A. and Ohtsuki T., "Performance analysis of BP-based algorithms for irregular low-density parity-check codes on fast Rayleigh fading channel", IEEE 60th Vehicular Technology Conference, 2004. VTC2004-Fall. vol. 4,pp.2530-2534, 2004.

[8]   Farid Ghani, Abid Yahya and Abdul Kader, "New Qc-LDPC Codes Implementation on FPGA platform in Rayleigh Fading Environment", IEEE Symposium on Computer & Information, 2011.

[9]   Savin V., "Min-Max decoding for non-binary LDPC codes", in Procedure IEEE International Symposium on Information Theory, ISIT, Pages 960 - 964, 2008.

[10]   Lin Jun, Sha Jin, Wang Zhongfeng, Li Li, "An Efficient VLSI Architecture for Nonbinary LDPC Decoders", IEEE Transactions on Circuits and Systems II: Express Briefs, Volume: 57 , Issue: 1, Pages 51 - 55, Jan. 2010.

[11]   Lin Jun, Sha Jin, Li Li, "Efficient decoder design for nonbinary quasi-cyclic LDPC codes", IEEE Transition on Circuits System I, Regular Papers, Volume 57, No. 5, Pages 1071-1082, May 2010.

[12]   X. Zhang and F. Cai, ``Efficient partial -parallel decoder architecture for Quasi-cyclic nonbinary LDPC codes", IEEE Transactions on Circuits and Systems I, Regular Papers, Volume 58, No.2, Pages 402-414, Feb 2011.

[13]   Zhang X. and Cai F., "Reduced-complexity decoder architecture for nonbinary LDPC codes", IEEE Transition of Very Large Scale Integr. (VLSI) System, volumn 19, no. 7, Pages 1229-1238, July 2011.

[14]   He Kai, Sha Jin, Wang, Zhongfeng, "Nonbinary LDPC Code Decoder Architecture With Efficient Check Node Processing", IEEE Transactions on Circuits and Systems II: ExpressLee, S.hyun. & Kim Mi Na, (2008) "This is my paper", ABC *Transactions on ECE*, Vol. 10, No. 5, pp120-122.

16

Performance Analysis Of FLS, EXP, LOG And M-LWDF Packet Scheduling Algorithms In Downlink 3GPP LTE System

Farhana Afroz[1], Shouman Barua[2], Kumbesan Sandrasegaran[2]

Faculty of Engineering and Information Technology,
University of Technology, Sydney, Australia

ABSTRACT

Long-Term Evolution (LTE), an emerging and promising fourth generation mobile technology, is expected to offer ubiquitous broadband access to the mobile subscribers. In this paper, the performance of Frame Level Scheduler (FLS), Exponential (EXP) rule, Logarithmic (LOG) rule and Maximum-Largest Weighted Delay First (M-LWDF) packet scheduling algorithms has been studied in the downlink 3GPP LTE cellular network. To this aim, a single cell with interference scenario has been considered. The performance evaluation is made by varying the number of UEs ranging from 10 to 50 (Case 1) and user speed in the range of [3, 120] km/h (Case 2). Results show that while the number of UEs and user speed increases, the performance of the considered scheduling schemes degrades and in both case FLS outperforms other three schemes in terms of several performance indexes such as average throughput, packet loss ratio (PLR), packet delay and fairness index.

KEYWORDS

LTE, packet scheduling, QoS, FLS, M-LWDF, EXP rule, LOG rule

1. INTRODUCTION

The continuously increasing demand of real-time (RT) multimedia services along with high speed internet access and the need of having ubiquitous access to them even in high mobility scenarios are acting as a driver toward the evolution of wireless cellular networks. To keep pace with this rising demand, the Third-Generation Partnership Project (3GPP) introduced LTE which is also marketed as 4G mobile network. LTE network targets to provide high peak data rates (100 Mbps in downlink and 50 Mbps in uplink within 20 MHz bandwidth), spectrum flexibility (1.25 to 20 MHz), improved system capacity and coverage, low user-plane latency (less than 5 ms), high spectral efficiency, support of wide user mobility, reduced operating cost, enhanced support for end-to-end Quality of Service (QoS) and seamless interoperability with existing systems [1, 2].

In this context, effective utilization of radio resources becomes crucial. LTE radio access network (also known as E-UTRAN, Evolved-UMTS Terrestrial Radio Access Network) uses OFDMA (Orthogonal Frequency Division Multiple Access) radio access technology in downlink in which the available bandwidth is divided into parallel narrow-band orthogonal subcarriers with sub-carrier spacing of 15 kHz irrespective of total bandwidth and each UE is allocated with a set of subcarriers depending on user's requirements, existing system load, and the configuration of system [3]. E-UTRAN consists of eNBs only (the LTE terminology for base station) where all RRM (Radio Resource Management) functions such as physical layer functions, scheduling, admission control etc. are performed. Packet scheduling is the process by which available radio resources are allocated among active users in order to (re)transmit their packets so as the QoS requirements of the users are satisfied [4]. The main objectives of packet scheduling are to maximize the cell capacity, to satisfy the minimum QoS needs for the connections, and to maintain adequate resources for best-effort users with no strict QoS requirements [5]. LTE packet scheduling mechanism is not specified by 3GPP, rather it is open for the vendors to implement their own algorithm. Different packet scheduling schemes has been proposed for LTE system. In this paper, the performance of FLS, LOG rule, EXP rule, and M-LWDF packet scheduling strategies has been studied by varying the number of users and users' speed.

The rest of this paper is organized as follows. A generalized packet scheduling model in the downlink LTE system is illustrated in section 2. Section 3 summarizes the dynamic packet scheduling schemes which were used in simulations followed by descriptions of the simulation scenarios and simulation results in section 4. Finally, section 5 concludes the paper.

2. DOWNLINK PACKET SCHEDULING MODEL

In downlink LTE system, the smallest unit of radio resource that can be allocated to a user for data transmission is known as Physical Resource Block (PRB) which is defined both in time and frequency domain [5]. In the frequency domain, the total available bandwidth is split into 180 kHz sub-channels, each sub-channel corresponds to 12 consecutive and equally spaced subcarriers with sub-carrier spacing of 15kHz (i.e. each sub-channel is of 12×15 =180kHz). In the time domain, the time is divided into frames and each LTE frame contains 10 consecutive TTIs (Transmission Time Interval). Each TTI is of 1ms duration and consists of two time slots, each of 0.5ms duration. Each time slot corresponds to 7 OFDM symbols (with short cyclic prefix). Resource allocation is performed on TTI basis. A time/frequency radio resource that spans over one time slot of 0.5ms in the time domain and one sub-channel (180 KHz) of 12 subcarriers in the frequency domain is known as Resource Block (RB). On every TTI, the RB pairs (in time domain) are allocated to a UE for data transmission.

The downlink packet scheduler aims to dynamically determine to which UE(s) to transmit packets and for each of the selected UE(s), on which Resource Block(s) (RB) the UE's Downlink Shared Channel (DL-SCH) will be transmitted [6]. A simplified packet scheduler model in LTE downlink system is shown in Fig. 1. In every TTI, each UE sends its CQI (Channel Quality Indicator) report computed from the downlink instantaneous channel condition to the serving eNB. At eNB, a buffer is assigned for each UE. Packets arriving at the buffer are time stamped and queued for transmission as FIFO (First In First Out) basis. On every TTI, scheduling decision takes place based on packet scheduling algorithms and one or more PRBs can be scheduled for each UE. There are specific scheduling criteria (e.g. channel condition, traffic type, head of line (HOL) packet delay, queue status etc.) for different scheduling strategies and depending on the scheduling criteria, users are prioritized. On each PRB, eNB choose a user with highest metric to transmit its packets. Once a user is selected, the number of bits transmitted per PRB depends on assigned Modulation and Coding Scheme (MCS) [7, 8].

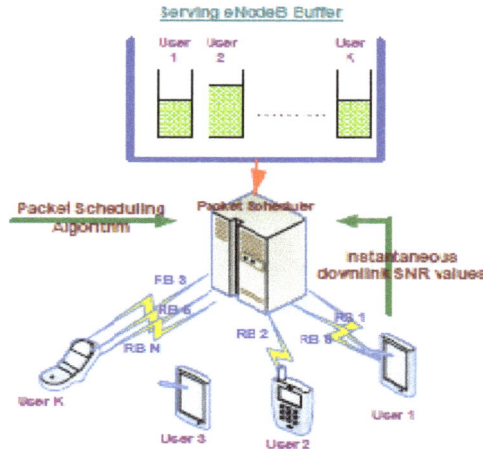

Fig. 1: A general LTE downlink packet scheduling model [8]

3. PACKET SCHEDULING STRATEGIES

LTE packet scheduling algorithm aims to maximize system performance. Different scheduling schemes have been proposed to support real-time (RT) and non real-time (NRT) applications. In this section, the algorithms that are considered in this paper will be described.

3.1. Maximum-Largest Weighted Delay First (M-LWDF)

M-LWDF [9] algorithm was proposed to support multiple real-time data users with different QoS requirements in CDMA-HDR system. A user is scheduled based on the following priority metric, M.

$$M = argmax \, a_i W_i(t) \frac{R_i(t)}{\bar{R}_i(t)} \tag{1}$$

$$\text{and } a_i = -\frac{log \delta_i}{\tau_i} \tag{2}$$

where $W_i(t)$ is the HOL packet delay of user i at time t, τ_i is the delay threshold of user i and $\square_i$ denotes the maximum probability of HOL packet delay of user i to exceed the delay threshold of user i.

Since, this scheme considers HOL packet delay together with PF properties, good throughput and fairness performance with a relatively low packet loss ratio (PLR) can be achieved using this algorithm.

3.2. Frame Level Scheduler (FLS)

This QoS (Quality of Service) aware packet scheduling algorithm was proposed in [10] for RT downlink communications. FLS is a two-level scheduling strategy where the two distinct levels (upper level and lower level) interact with each other to dynamically allocate RBs to the users. At upper level, a resource allocation scheme (namely FLS), which utilizes a D-T (Discrete-Time) linear control loop, is implemented. FLS specifies the amount of data packets that a RT source should transmit frame by frame to satisfy its delay constraint. At lower level, in every TTI, RBs are allocated to the UEs using Proportional Fair (proposed in [11]) scheme with taking into consideration the bandwidth requirements of FLS. Particularly, the scheduler at the lower layer defines the number of TTIs/RBs through which each RT source will send its data packets. The amount of data to be transmitted is given by the following equation:

$$v_i(k) = h_i(k) * q_i(k) \tag{3}$$

Where, $v_i(k)$ is the amount of data to be transmitted by the i-th flow in k-th LTE frame, "*" is the D-T convolution operator, $q_i(k)$ is the queue level. The above equation says that $v_i(k)$ is obtained by filtering the signal $q_i(k)$ through a time-invariant linear filter with pulse response $h_i(k)$.

3.3. Exponential (EXP) Rule

The Exponential rule [12], a channel aware/QoS aware scheduling strategy, was proposed to offer Quality of Service (QoS) guarantees to the users over a shared wireless link. It explicitly considers the channel conditions and the state of the queues while making scheduling decisions. The following two rules are called EXP rule.

The Exponential (Queue length) rule (EXP-Q) selects a single queue for service in time slot t

$$i \in i(S(t)) = argmax_i \, \gamma_i \, \mu_i(t) \exp \left(\frac{a_i Q_i(t)}{\beta + [\bar{Q}(t)]^\eta} \right) \tag{4}$$

where $\mu_i(t) \equiv \mu_i^{m(t)}$ and $\bar{Q}(t) \doteq \left(\frac{1}{N} \right) \Sigma_i \, a_i Q_i(t)$

Likewise, the Exponential (Waiting time) rule (EXP-W) selects for service a queue

$$i \in i(S(t)) = argmax_i \, \gamma_i \, \mu_i(t) \exp \left(\frac{a_i W_i(t)}{\beta + [\bar{W}(t)]^\eta} \right) \tag{5}$$

where $\bar{W}(t) \doteq \left(\frac{1}{N} \right) \Sigma_i \, a_i W_i(t)$

Here, $\gamma_1, \dots. \gamma_N$ and $a_1 \dots. a_N$ are arbitrary set of positive constants, $\eta \in (0,1)$ is fixed and β is positive constant. The EXP rule chooses either EXP-W or EXP-Q rule for service a queue.

3.4. LOG Rule

This channel aware/QoS aware strategy was designed to give a balanced QoS metrics in terms of robustness and mean delay [13]. Similar to the EXP rule, the scheduler allocates service to the user in a manner that maximizes current system throughput, with considering that traffic arrival and channel statistics are known. When users' queues are in state q and the channel spectral efficiencies of them are $K \equiv (K_i : 1 \leq i \leq N)$, LOG rule scheduler serves a user i_{LOG}:

$$i_{LOG}(q, K) \in arg \, max_{1 \leq i \leq N} \, b_i \log (c + a_i Q_i) \times K_i \tag{6}$$

Here, b_i, a_i, c are fixed positive constants, $0 < \eta < 1$ and Q_i represents the queue length.

4. PERFORMANCE EVALUATION

The performance evaluation of FLS, EXP rule, LOG rule and M-LWDF scheduling schemes with increasing number of UEs (Case 1) and varying UE's speed (Case 2) will be reported in this section. To this aim, an open source simulator namely LTE-Sim [14] has been adopted. LTE-Sim simulator exploits Jain's fairness method [15] to calculate fairness index among UEs. The propagation loss model includes the following:

-Fast fading: Jakes model

-Path loss: $L=128.1+37.6\log10d$ @2GHz,
where d is the distance between user and eNB in Km
-Penetration loss: 10dB
-Shadow fading: Lognormal distribution with mean 0 and standard deviation 8dB

4.1. Case 1: Effects of number of users

The performance of FLS, EXP rule, LOG rule, and M-LWDF downlink packet scheduling schemes with increasing the number of UEs is analyzed herein. For multimedia flows, the considered scheduling schemes have been compared based on several performance metrics named average throughput, PLR, delay, and the fairness index. For best effort (BE) flows, since there is no strict QoS requirements, a comparison among these scheduling strategies is reported on the basis of average throughput only.

4.1.1. Simulation scenario

A single urban macro cell with interference simulation scenario with each UE having single flow (video or VoIP or BE) and 40% UEs receiving video flows, 40% users receiving VoIP flows and the rest 20% receiving BE flows has been taken into consideration to study the effects of number of users on the performance of the scheduling strategies described above. A number of UEs ranging from 10 to 50 are uniformly distributed and moving with a speed of 120 km/h in random direction within a cell. Table 1 shows the simulation parameters.

Table 1. Simulation parameters

Parameters	Value
Simulation time	150 sec
Cell radius	1 Km
User speed	120 km/h
Video bit rate	242 kbps
Frame structure	FDD
Bandwidth	10 MHz
Flow duration	120 sec
Maximum delay	0.1 sec

4.1.2. Results and Discussion

The average throughput graphs of video, VoIP and best effort flows in Fig. 2 demonstrate that the average throughput degrades while the number of users increases and FLS algorithms shows best average throughput performance for multimedia flows . As seen in Fig. 2(a), the average throughput of video flow falls upon increasing number of users for all the considered scheduling algorithms. For FLS algorithm, while the number of users increases from 10 to 20, the average throughput sharply falls followed by a steady decline in average throughput when the cell is charged with more than 20 users. M-LWDF and LOG rule provides almost identical throughput performance and EXP rule shows higher average throughput than these two schemes. The average throughput per VoIP flow (shown in Fig. 2(b)) maintains almost the constant level at 3000 bps in the user range of 10 to 40 for all four schemes. When the user number exceeds 40, the average throughput slowly drops for all four schemes with increasing users. These no-variation trend of VoIP average throughput may be due to the VoIP traffic model (ON/OFF Markov chain) and the ON/OFF periods used during simulation. The average throughput graph of best effort flow in Fig. 2(c) depicts that while the user number increases, LOG rule and M-LWDF provide better average throughput performance compared with FLS algorithm whereas, EXP-rule provides higher average throughput than FLS scheme for the users ranging from 20 to 50.

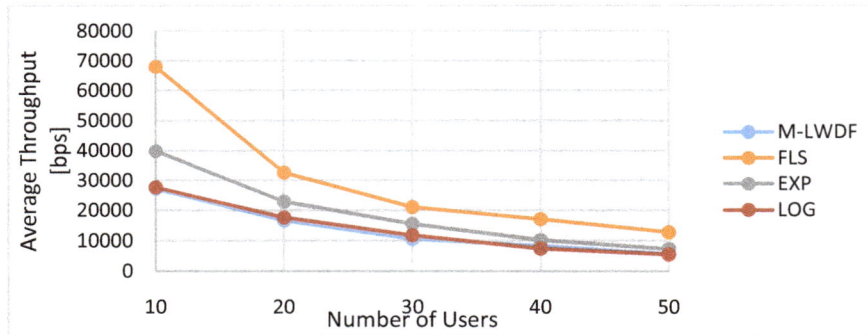

(a)

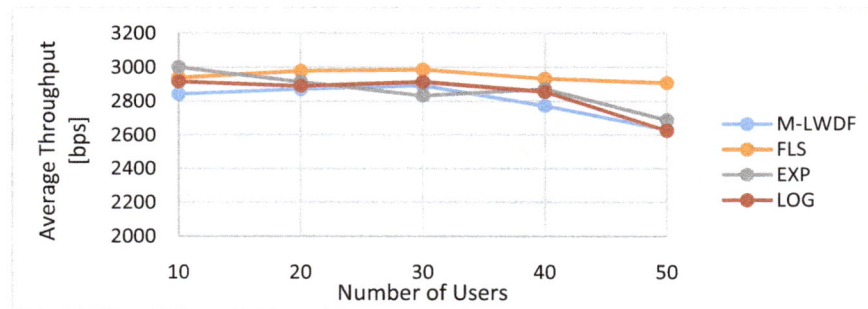

(b)

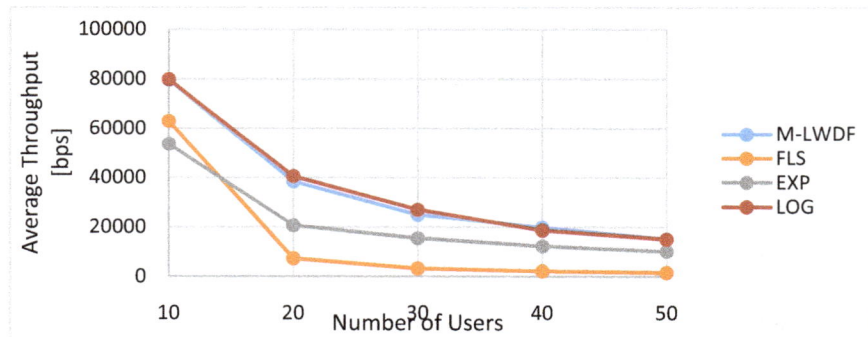

(c)

Fig. 2: Average throughput of (a) video flow (b) VoIP flow (c) BE flow

Fig. 3, showing the packet loss ratio (PLR) experienced by video and VoIP flows, describes that the PLR increases with increasing number of users because of increased network loads and the PLRs experienced by VoIP flows are considerably smaller than that of video flows for all four scheduling schemes. It can be also noticed that for multimedia flows, lowest PLRs are achieved using FLS algorithm and EXP rule offers better performance (i.e. smaller PLR) as compared with LOG rule and M-LWDF. As seen in Fig. 3(a), for video flow, LOG rule and M-LWDF provide almost same PLR performance. From Fig. 3(b), it is noticed that for VoIP flow, FLS algorithm maintains below 1% of PLR in the user range of 10 to 50. The PLRs remain within 5% for LOG rule and M-LWDF scheme and within 3% for EXP rule in the range of 10-40 users.

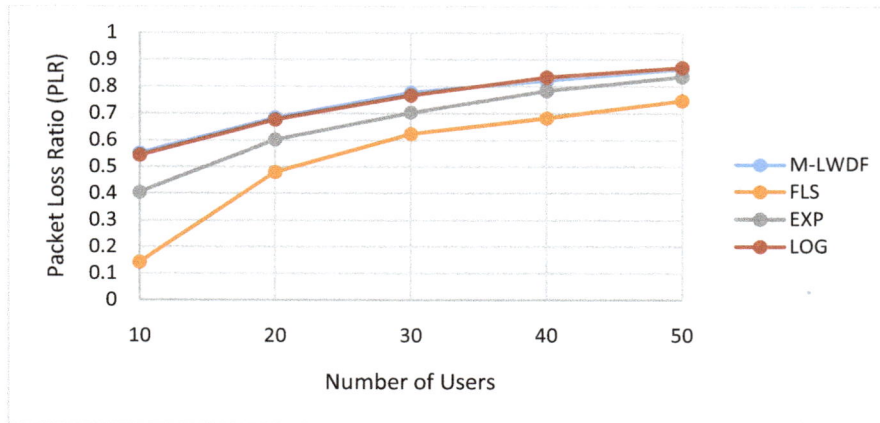

(a)

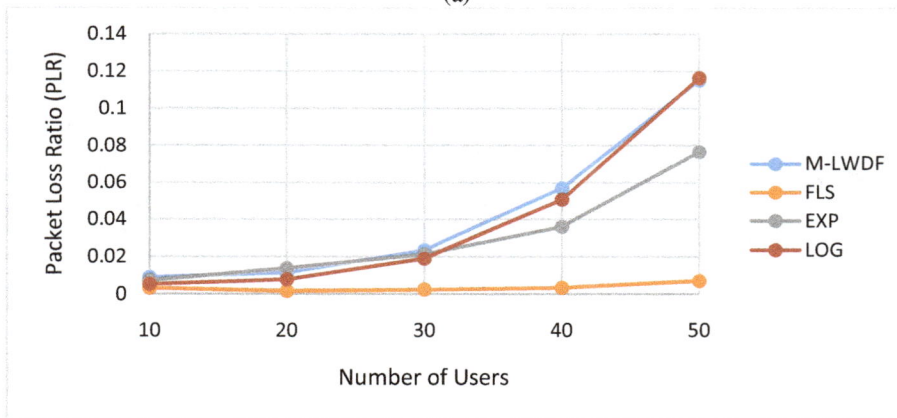

(b)

Fig. 3: PLR of (a) video flow (2) VoIP flow

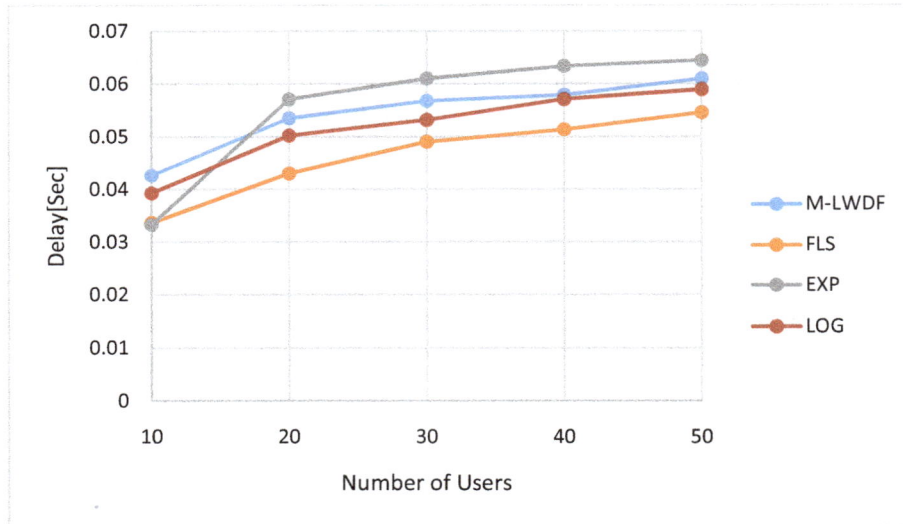

(a)

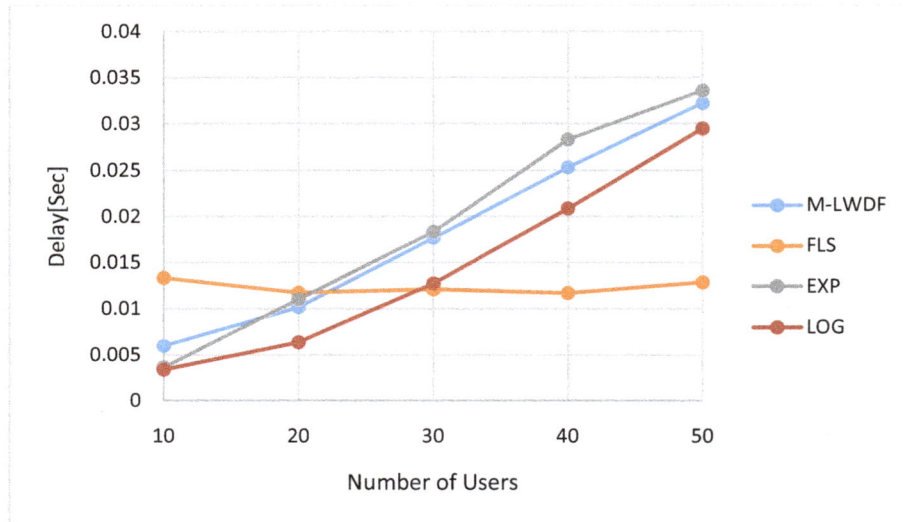

(b)

Fig. 4: Packet delay of (a) video flow (b) VoIP flow

As seen in Fig. 4(a), the packet delay of video flow gradually increases with increasing number of users for all four schemes and FLS is showing lowest delay among them. Fig. 4(b) showing the packet delay of VoIP flow illustrates that, for FLS scheme the packet delay maintains almost same level while increasing number of users. It is observed that FLS is giving lowest upper bound of the delay among four schemes and hence shows the lowest PLR.

Fig. 5(a) illustrates that for video flow, fairness index degrades with increasing number of users for all the four algorithms and FLS scheme ensures highest degree of fairness among them. In case of VoIP flow (Fig. 5(b)), fairness indexes are maximum when the cell is charged with 10 users and minimum when the user number is 50 for all four scheduling schemes with FLS is having the highest fairness index.

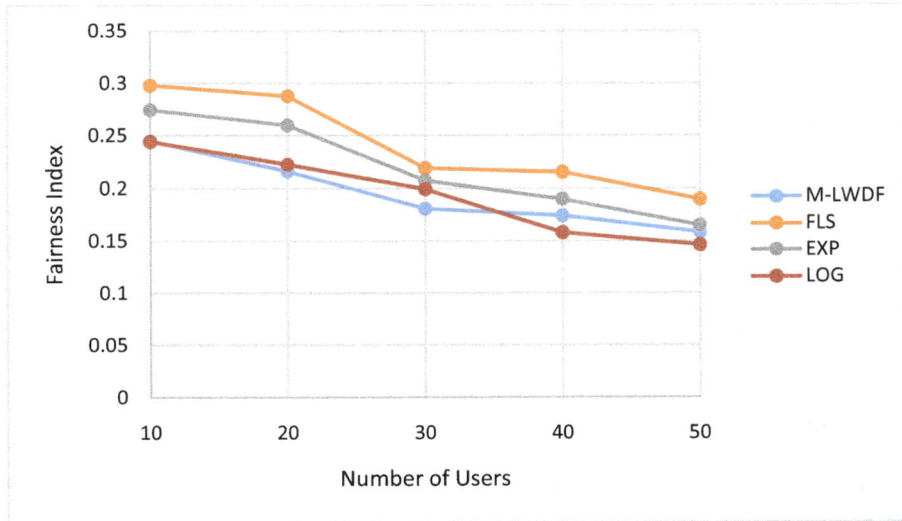

(a)

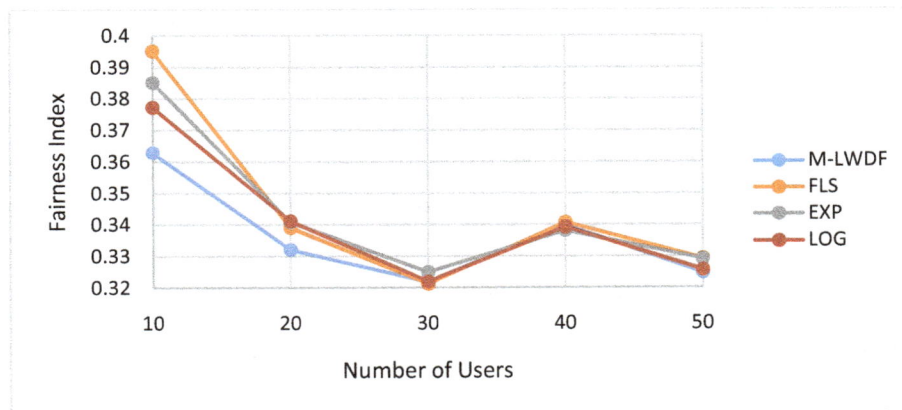

(b)

Fig. 5: Fairness index of (a) video flow (b) VoIP flow

4.2. Case 2: Effects of users' speed

In this part, two distinct user speed (pedestrian speed – 3 km/h and vehicular speed – 120 km/h) are considered to study the effects of user's speed on the performance of the FLS, EXP rule, LOG rule and M-LWDF packet scheduling algorithms.

4.2.1. Simulation scenario

The simulation scenario considered here is identical to that of Case 1 (Subsection 4.1.1). The simulation parameters are given in Table 2.

Table 2. Simulation parameters

Parameters	Value
Simulation time	150 sec
Cell radius	1 Km
User speed	3 km/h and 120 km/h
Video bit rate	242 kbps
Frame structure	FDD
Bandwidth	10 MHz
Flow duration	120 sec
Maximum delay	0.1 sec

4.2.2. Results and Discussion

Fig. 6 illustrates the effects of user speed on the average throughputs of BE flow, video flow and VoIP flows. As seen, the average throughputs of video flow (as seen in Fig. 6(a)) and BE flow (Fig. 6(c)) decrease with increasing users' speed from 3 km/h to 120 km/h for all four schemes. It is expected that average throughput decrease with increasing user speed because at higher speed channel quality measured by UE becomes worse, which in turn triggers lower order modulation to be selected and thus results in lower average throughput. From the graph of VoIP average throughput (Fig. 6(b)), it is observed that for FLS, the average throughputs of VoIP flow maintains almost the same level while the user speed increases. For EXP rule, LOG rule, M-LWDF, the VoIP average throughput degrades with increasing user speed at higher speed. The packet loss ratios (PLRs) of video flow and VoIP flow, reported in Fig 7(a) and 7(b) respectively, show that for multimedia flows, the PLRs become greater when the users are at higher speed. The reason is- at higher speed poor link adaptation occurs. As seen in Fig. 8, the packet delay increases with increasing user speed for all four schemes. Fig. 9(a) demonstrates that, for video flow the fairness index falls at higher user speed for all four algorithms and FLS provides higher degree of fairness at both user speed. It is seen from the Fig. 9(b) that for VoIP flow, the considered scheduling schemes provide approximately same fairness index irrespective of user speed.

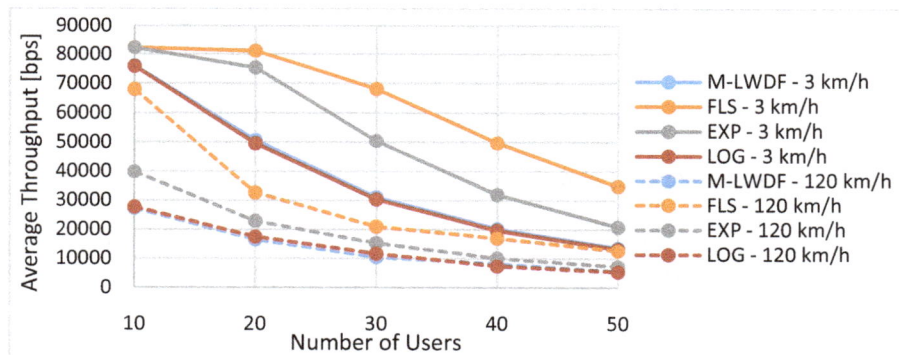

(a)

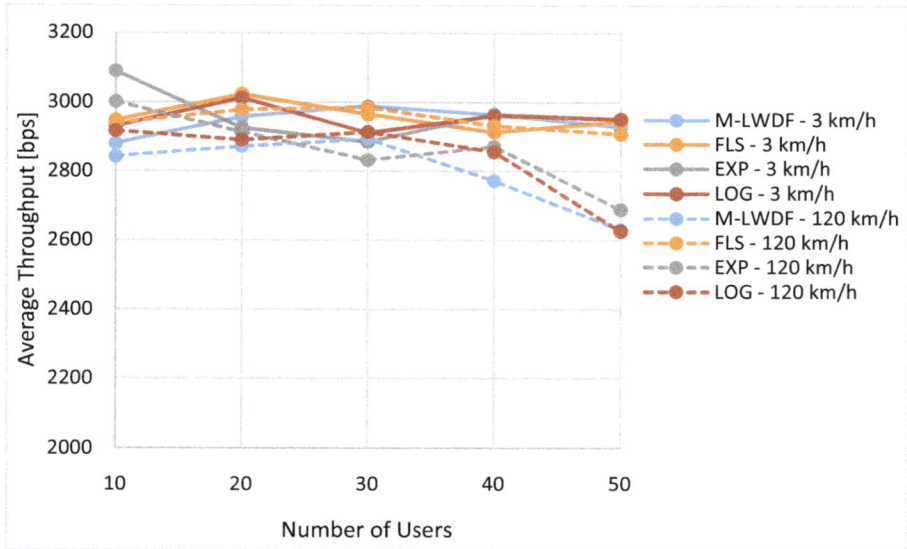

(b)

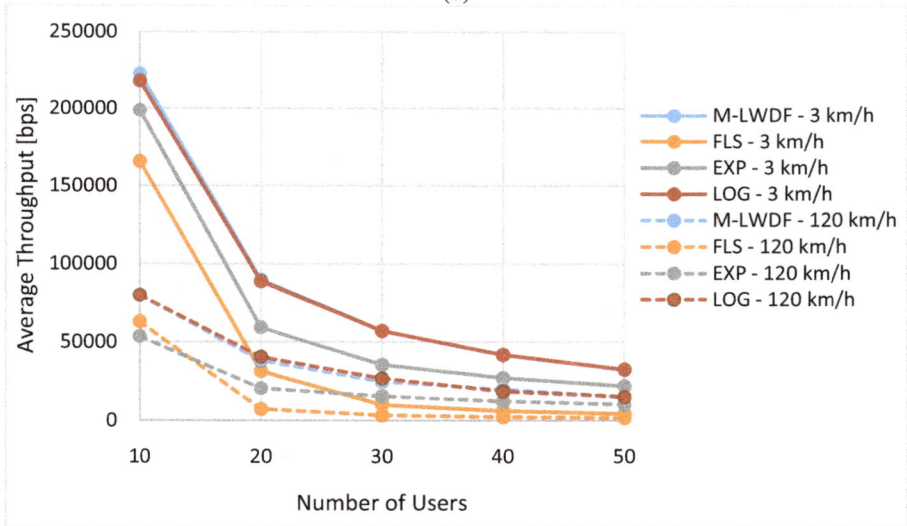

(c)

Fig. 6: Average throughput of (a) video flow (b) VoIP flow (c) BE flow

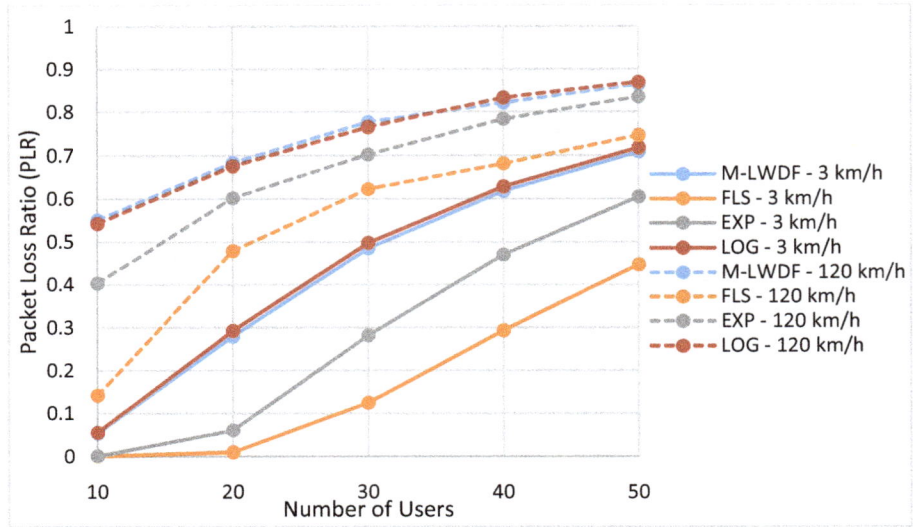

(a)

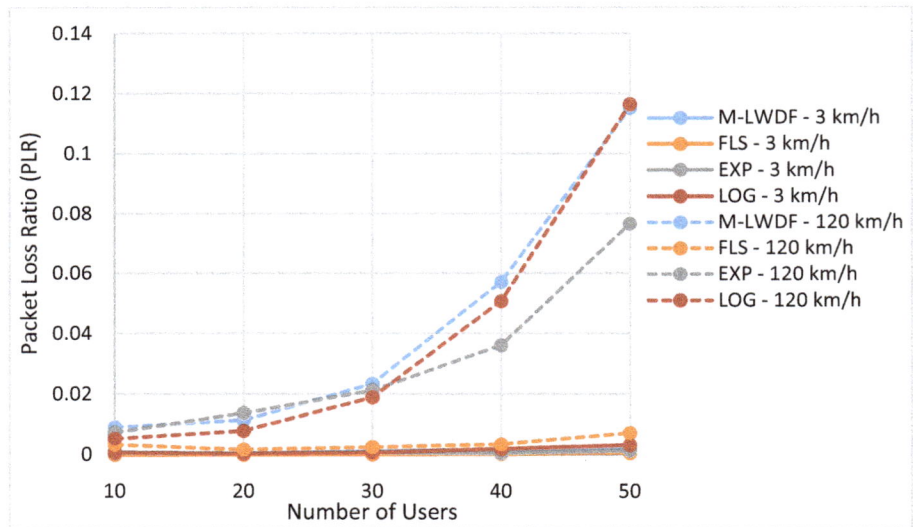

(b)

Fig. 7: PLR of (a) video flow (b) VoIP flow

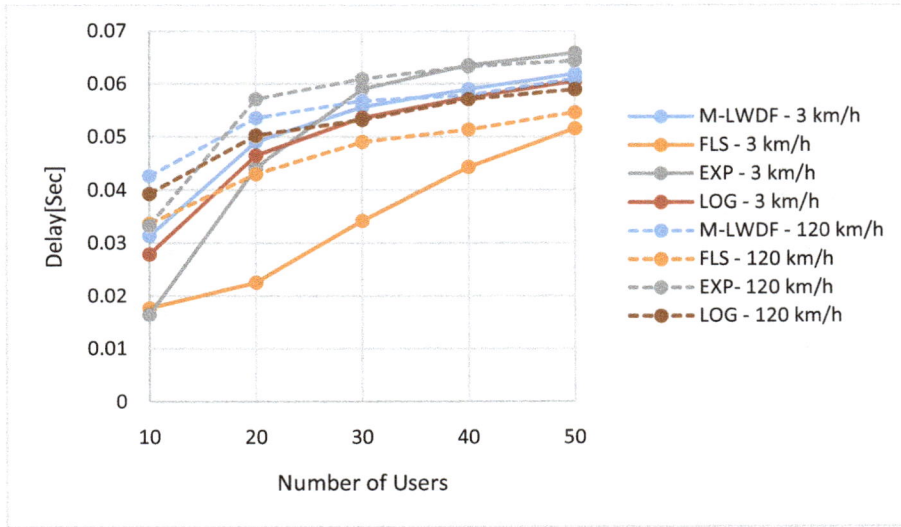

(a)

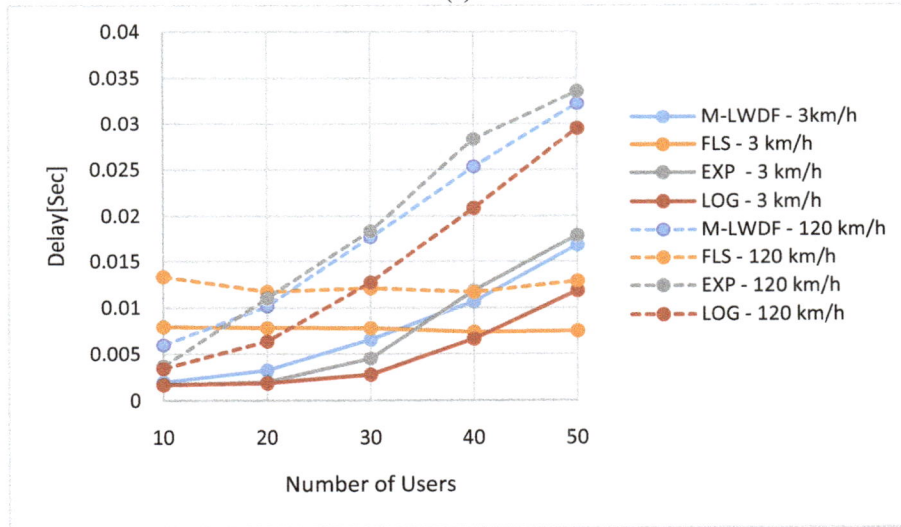

(b)

Fig. 8: Packet delay of (a) video flow (b) VoIP flow

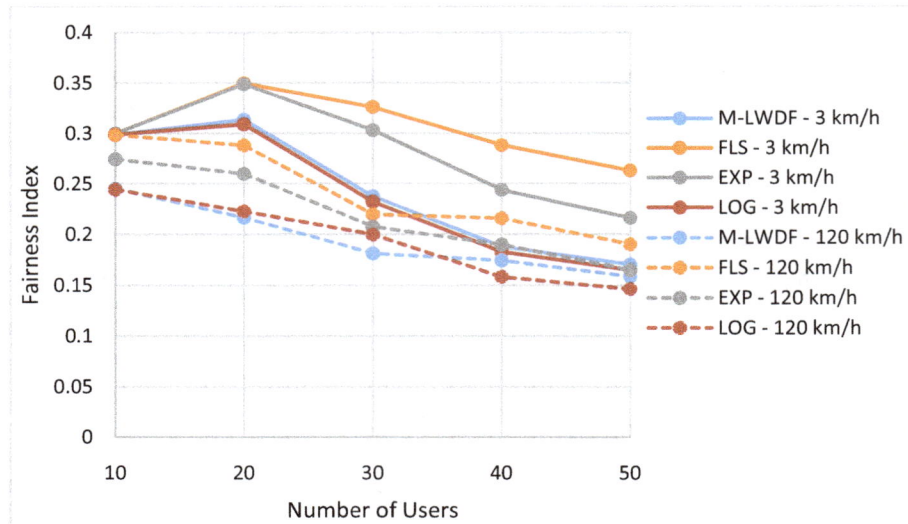

(a)

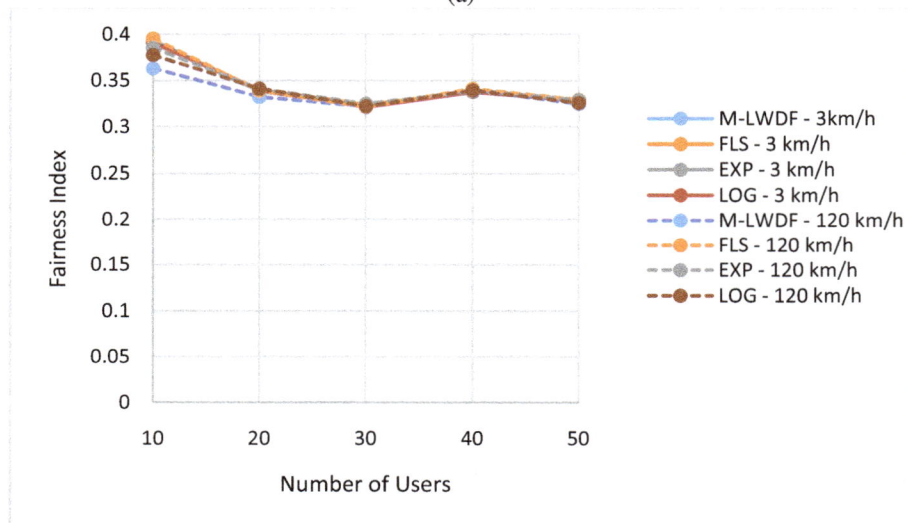

(b)

Fig. 9: Fairness index of (a) video flow (b) VoIP flow

5. CONCLUSION

In this paper, the performance study of FLS, EXP rule, LOG rule and M-LWDF packet scheduling algorithms in LTE downlink has been performed while varying number of users and users' speed. The simulation results show that overall FLS scheme outperforms other three schemes in terms of average throughput, PLR, delay, and fairness index. It is also reported that the performance of simulated packet scheduling strategies drops noticeably while the users' speed increases. Our future work includes to simulate and compare the performance of LTE downlink packet scheduling algorithms with different scenarios.

REFERENCES

[1]   A. Ghosh and R. Ratasuk, Essentials of LTE and LTE-A: Cambridge University Press, 2011.

[2]   3GPP Technical Report, TR 25.913, "Requirements for Evolved UTRA (E-UTRA) and Evolved UTRAN (E-UTRAN)," version 7.0.0, June 2005.

[3]   H.A.M. Ramli, K. Sandrasegaran, R. Basukala, L. Wu, "Modeling and Simulation of Packet Scheduling in the Downlink Long Term   Evolution System," 15th Asia-Pacific Conference on Communications, Oct. 2009, pp.68-71.

[4]   G. Piro, L. A. Grieco, G. Boggia, F. Capozzi, and P. Camarda, "Simulating LTE cellular systems: an open- source framework," IEEE Transactions on Vehicular Technology, Feb. 2011, vol. 60, pp. 498-513.

[5]   H. Holma and A. Toskala, LTE for UMTS : OFDMA and SC-FDMA based radio access, Chichester, U.K.: Wiley, 2009.

[6]   E. Dahlman, S. Parkvall, and J. Sköld, 4G LTE/LTE-Advanced for Mobile Broadband: Academic Press, 2011.

[7]   R. Basukala, H. A. M. Ramli, and K. Sandrasegaran, "Performance analysis of EXP/PF and M-LWDF in downlink 3GPP LTE system," 1st AH-ICI on internet, Nov. 2009, pp. 1–5.

[8]   H. A. M. Ramli, R. Basukala, K. Sandrasegaran, and R. Patachaianand,"Performance of well known packet scheduling algorithms in the downlink 3GPP LTE system," in Communications (MICC), IEEE 9th Malaysia International Conference , Dec. 2009, pp. 815-820.

[9]M. Andrews, K. Kumaran, K. Ramanan, A. Stolyar, P. Whiting, and R. Vijayakumar, "Providing Quality of Service over a Shared Wireless Link," IEEE Communications Magazine, vol. 39, pp. 150-154, 2001.

[10] Giuseppe Piro, Luigi Alfredo Grieco, Gennaro Boggia, Rossella Fortuna, and Pietro Camarda, (2011) "Two-Level Downlink Scheduling for Real-Time Multimedia Services in LTE Networks", IEEE Transaction on Multimedia, Vol. 13, No. 5.

[11] A. Jalali, R. Padovani, and R. Pankaj, "Data Throughput CDMAHDR a High Efficiency-High Data Rate   Personal Communication Wireless System," in IEEE 51st Vehicular Technology Conference Proceedings, Tokyo, 2000, pp. 1854-1858.

[12] S. Shakkottai and A. Stolyar, (2002) "Scheduling for Multiple Flows Sharing a Time-Varying Channel: The Exponential Rule", Analytic Methods in Applied Probability, Vol. 207 of American Mathematical Society Translations, Series 2, A Volume in Memory of F. Karpelevich, pp. 185202, American Mathematical Society, Providence, RI, USA.

[13] B. Sadiq, S.J.Baek, and G. de Veciana, (2009) "Delay-Optimal Opportunistic Scheduling And Approximations: The Log rule", Proceedings of the 27th Annual Joint Conference on the IEEE Computer and Communications Societies (INFOCOM 09), pp. 19.

[14] G. Piro, L. A. Grieco, G. Boggia, F. Capozzi and P. Camarda, (2011) "Simulating LTE Cellular Systems: An Open-Source Framework", IEEE Transaction on Vehicular Technology, Vol. 60, No. 2.

[15] R. Jain, D. Chiu, and W. Hawe, A quantitative measure of fairness and discrimination for resource allocation in shared computer systems, Digital Equip. Corp., Littleton, MA, DEC Rep., DEC-TR-301, Sep. 1984.

17

MULTI-PATH ROUTING AND CHANNEL ASSIGNMENT FRAMEWORK FOR MESH COGNITIVE RADIO NETWORK (MRCAMC)

[1]Amjad Ali, [2]Muddesar Iqbal, [1]Adeel Baig, [3]Xingheng Wang

[1]School of Electrical Engineering and Computer Sciences National University of Science and Technology, Pakistan
{amjad.ali,adeel.baig}@seecs.edu.pk
[2]Faculty of Computer Science & Information Technology, University of Gujrat, Pakistan
m.iqbal@uog.edu.pk
[3]College of Engineering, Swansea University, Swansea, UK
xingheng.wang@swansea.ac.uk

ABSTRACT

Dynamic spectrum access is an attractive area of research these days. Cognitive Radio (CR) enabled networks are being deployed to effectively utilize the RF spectrum. Wireless mesh networks have been experiencing the bandwidth scarcity but such networks can easily enhance their throughput by using the CR transceivers as these networks have the capability of multipath routing. Existing routing proposals for Mesh Cognitive Radio Networks (MCRNs) are not considering the dynamic spectrum availability. They try to treat CRNs as traditional wireless networks. Even some proposals treat the problems of these networks like that of wired networks and use the same parameter for route discovery as for wired networks. In this paper, we propose a joint interaction between on-demand routing and channel assignment that accounts the characteristics of CRNs.

KEYWORDS

CRNs, Routing, Mesh Cognitive Radio Networks

1.INTRODUCTION

Wireless communication has established itself as a popular access technology due to the user preference for flexibility, but its static channel allocation based spectrum management scheme is still a main problem. This static channel allocation scheme is very inefficient as it only allows the licensed users, Primary Users (PUs), to access the channels and does not permit unlicensed users, Secondary Users (SUs), to access the channels although the channels are idle or underutilized. This leads to the wastage of the spectrum resources. To efficiently utilize radio spectrum resources, a novel communication paradigm known as Cognitive Radio (CR) or Dynamic-Spectrum-Access (DSA) has been proposed[1] [2] [3]. Cognitive Radio Networks

(CRNs) utilize the available spectrum opportunistically in the locality in which they operate. CR technology minimizes the wastage of radio spectrum band.

Wireless Mesh Networks (WMNs) [4] suffer from bandwidth scarcity. However, multipath routing can easily be achieved in WMNs. Thus splitting the flow among the multiple available paths can make simultaneous transmission to reduce the overall delay. Performance of WMNs can be elevated by using the CRs. Routing in Mesh Cognitive Radio Networks (MCRNs) [5] is a challenging task due to the diversity in the available channel set, data rates and reliability of the intermediate nodes. Without considering these issues routing cannot be useful in MCRNs as a large portion of useful time is utilized for routes discovery and route maintenance. Thus the critical issue in the MCRNs is the detection and avoidance of the PU interference as it affects the available channel set that further affects the route availability.

In MCRNs the existence of node itself has no significant meaning if the node has no common channel for communication with the neighboring nodes. If the node has some common channel with its neighbors for communications then we say node plays a key role in the network. Based on this argument we say the node/radio is virtual entity and channel/RF spectrum is physical entity. Thus deciding the route based on the physical location of the nodes/radios has no worth in MCRNs. Thus on demand routing technique is preferred for CRNs due to its dynamic nature. On demand protocol discovers the route from source to destination when it is required by source node instead maintaining the complete routing table for entire topology and updating it periodically. The mostly used on demand routing protocols are Dynamic Source Routing (DSR) [6] and Ad hoc on-demand Distance Vector (AODV) [7].

To the best of our knowledge, available routing proposals for the MCRNs select the intermediate nodes and channel on them heuristically without considering the dynamics of the CR. In this paper, we have proposed a new routing protocol for MCRNs that considers the dynamic of the CR. The rest of this paper is organized as follows. In Section 2 we discuss the related work and our motivation. In section3 we present in detail the proposed routing protocol with network modeling. Paper conclusion and future work are discussed in section 4.

2. RELATED WORK

Research on the CRNs has started since last few years and researchers all over the world are exploring the new ways to deploy the CRNs such that the RF resources could be efficiently utilized. The routing proposals available for Mobile Ad hoc Networks (MANETs) and wireless mesh networks could not be directly applied for CRNs as they do not consider the dynamics of CRNs.

SMR [8] is on demand multipath routing protocol for MANETs. It discovers multiple disjoint paths between source and destination pair and splits the data traffic on multiple established paths. SMR selects the best route based on the minimum hop count. Furthermore it considers only single channel and node level disjointness while studies show that spectrum wise disjointness is more common in CRNs equipped with multiple interfaces [8].

Multi-Flow Real-Time Transport Protocol "MRTP" [9] is another multipath routing protocol for mesh-based MANETs. MRTP is based on the Real Time Protocol (RTP) and specifically used for multicast application.

HC-IPSAG [10] is a cluster based routing proposal for CRN's. It splits the CRN's into clusters and each cluster is represented by head node. Each head node runs its own IPSAG protocol same like BGP.

Geographically based routing proposal for CRN's is proposed in [11], is derived from multi-hop, multi-channel adhoc and mesh networks. The proposed routing protocol focuses on most stable path that is indirectly achieved by simultaneous transmission over the multiple available channels and meeting the flow demand of CR node. It uses a novel routing metric based on probabilistic definition of available capacity over a channel.

In [12], a fuzzy based routing solution is proposed for CRN's. It enhances the throughput of CRN's by selecting the most stable channel when routes are being selected in addition to power factor that determines the amount of interference that primary user could afford. Channel stability is measured in term of channel utilization by primary user.

Multi-Radio Link-Quality Source Routing (MR-LQSR) [13] is multi-radio single path routing protocol for mesh-based MANETs. It uses link-state protocol for selecting the best path for source destination pair. Weighted cumulative expected transmission time (WCETT) is used for path selection. WCETT considers both link quality and minimum hop count.

SORP [14] and CARD [15] is single path routing protocols for MCRNs. Multipath Routing and Spectrum Access (MRSA) [16] is the only available multipath routing protocol for MCRNs. It is on demand and uses minimum hop count for selecting multiple paths between the source and destination pair. It handles the PU appearance on any path with the help of multiple paths used for same source destination pair. It does not consider the dynamics of CR while selecting the paths. Thus path selected by this routing protocol are not reliable while to ensure reliability is more important in CRNs than wired or other wireless networks.

3. NETWORK MODEL

Table 1: Index of symbols used in paper

Symbol	Description
$C_n(t)$	Actual channel n status at time slot t
$S_n(t)$	Sensing result of channel n at time t
δ_n, ε_n	Probability of errors in sensing results for channel n
α_n, β_n	Transition probabilities of channel states
$\mu_n(t)$	Primary user susceptibility at time slot for channel n
p_x^{tr}	Transmission probability of node x
θ_i	Collision range
E_{STX}	Successful transmission metric
$E_{RANK}^{x \leftarrow y}$	Rank of a link between node y and node x
$\mathfrak{N}$	Set of Radios

μ_{i,c_l}	Primary user susceptibility on link l for channel of path i
$c_{k,r}$	Candidate channel of node k for radio r
$^{\circ}C_{k,j}$	Common channels between node k and j
$\mathbb{R}_i$	Rank of path i

3.1. Primary Users Susceptibility

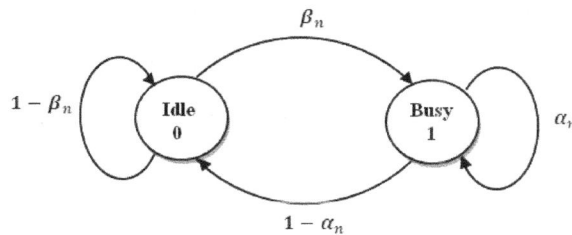

Figure1. Markov model for channel status

Let N= 1,1, (1,   2, ... , n)channels available in the whole spectrum band$^{\Psi}$. We assume that an individual node is only capable of accessing and sensing a subset of the whole spectrum, i.e. $\beta_i \subseteq \Psi$. let the channel availability for the secondary user be modeled by a Markov process given in Fig 1. To cater for the errors in the sensing process we can formulate two vectors, one consisting of the actual condition the channel is in and the other being the actual results of the sensing process. $S_n(t)$ denotes the sensing results achieved by a node and let $C_n(t)$ denote the actual channel state the channel was in. Since the channel can be in either of the two states, i.e. busy or idle, the errors induced in the sensing process and resulting in a false value of $S_n(t)$ can be represented by asymmetrical channel depicted in Fig. 2.

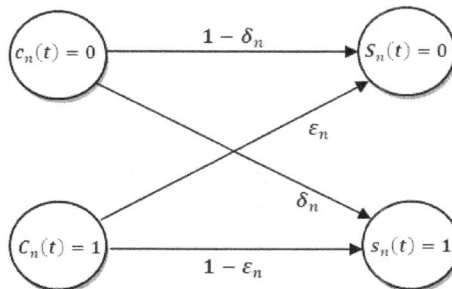

Figure2. Asymmetric channel for actual channel status and sensing results of node

The errors in the sensing process are defined by the following equations:

$$\text{Prob}(S_n(t) = 1 | C_n(t) = 0) = \delta_n \qquad (1)$$

$$\text{Prob}(S_n(t) = 0 | C_n(t) = 1) = \varepsilon_n \qquad (2)$$

The sensing results and the actual channel status are defined by a vector. The sensing vector is given by $\overrightarrow{S_n}(t) = [S_1(t), S_2(t), \ldots, S_n(t)]$ and the actual channel condition vector is given by $\overrightarrow{C_n}(t) = [C_1(t), C_2(t), \ldots, C_n(t)]$. We define primary user susceptibility for a channel as a belief of the channel being in busy state. The primary user susceptibility vector would then be given by $\overrightarrow{\mu_n}(t) = [\mu_1(t), \mu_2(t), \ldots, \mu_n(t)]$. Hence by utilizing Bayes law and conditional probability we can compute μ_n in the following way. The probability of channel being in the busy state, given some previous history of the channel ω_n is given by

$$\mu_n(t) = \text{Prob}(C_n(t) = 1 \mid \omega_n)$$

$$= \alpha_n \mu_n(t-1) + \beta_n \big(1 - \mu_n(t-1)\big)$$

$$= \pi_n(t) \qquad (3)$$

Equation (3) defines the primary user susceptibility as the channel being in the busy state in the previous time slot or having moved in the busy state from the idle state in the previous time slot. The history ω_n can be manipulated as:

1. If there is an acknowledgement received in time interval t, then in time t-1, the channel was busy in data transmission.
2. If there is a successful negotiation for transmission in the current slot then in the previous slot, the channel was busy in carrying requests.
3. If there is data received in time interval t then in t-1, the channel was busy in transmission.

Incorporating and conditioning on the channel sensing result S_n, we can write for correctly deciding upon the available channels as

$$\mu_n(t) = \text{Prob}(C_n(t) = 1 | S_n(t) = 1, \omega_n)$$

$$\mu_n(t) = \frac{\text{Prob}(S_n(t) = 1 | C_n(t) = 1, \omega_n)}{\sum_{I_n \in (0,1)} \text{Prob}(S_n(t) = 1 \mid C_n(t) = I_n, \omega_n)}$$

$$= \frac{\text{Prob}(S_n(t) = 1 | C_n(t) = 1) \text{Prob}(C_n(t) = 1 \mid \omega_n)}{\sum_{I_n} \text{Prob}(S_n(t) = 1 \mid C_n(t) = I_n) \text{Prob}(C_n(t) = I_n | \omega_n)}$$

$$\mu_n(t) = \frac{\pi_n(t)(1 - \varepsilon_n)}{\pi_n(t)(1 - \varepsilon_n) + [1 - \pi_n(t)]\delta_n} \qquad (4)$$

Where the value of $\pi_n(t)$ will be replaced as $\alpha_n \mu_n(t-1) + \beta_n(1 - \mu_n(t-1))$. Values of ε_n and δ_n can be computed by running the system on some time scale [0-T] and updated after a significant percentage change Δ occurs in the value, hence it is specific to a particular nodes operation. We have defined the probability $\mu_n(t) = \pi_n(t)$ as a function of previous time interval. Hence by performing a recursive operation we can write for any future time instant $t + \tau$, the probability value as

$$\mu_n(t+\tau) = \mu_n(t)(\alpha_n - \beta_n) + \beta_n \sum_{i=0}^{\tau-1} (\alpha_n - \beta_n)^i \quad \forall \tau > 0 \quad (5)$$

The $\mu_n(t+\tau)$ value is used in algorithm 2 for computing the channel status over a period of time. Hence if a node is unable to perform channel sensing at a particular slot, it can use the approximations from this equation.

Here we try to formulate a metric that can combine the effect of links on a particular hop while also providing information of the total hops traversed. A metric that gives proportional importance to the combined channel states will suit the purpose. We have already defined the probability with which a node is sure that a channel will not be susceptible to the primary user or simple the probability $\mu_n(t)$ with which the channel would be idle. We can define here the probability for unsuccessful transmission p between a node x and y in the form of busy channel and transmission probability. Let the transmission probability be given by p_x^{tr} for node x and p_y^{tr} for node y. Then for unsuccessful transmission on a bidirectional link basis p can be defined as

$$p = 1 - \left[1 - (\mu_n^x(t))(p_x^{tr})\right] \times \left[1 - (\mu_n^y(t))(p_y^{tr})\right] \quad (6)$$

Where $\mu_n^x(t)$ and $\mu_n^y(t)$ are the probabilities of miss detection, i.e. the channel was in busy state and the sensing result was a 0 (busy). Similar to equation (4), we can compute the probability of miss detection as

$$\mu_n^{x,y}(t) = Prob(C_n(t) = 1 | S_n(t) = 0, \omega_n)$$

$$\mu_n^{x,y}(t) = \frac{\pi_n(t)(\varepsilon_n)}{\pi_n(t)(\varepsilon_n) + [1 - \pi_n(t)](1 - \delta_n)} \quad (7)$$

The probability of transmission can be defined in terms of a collision range under which we allow a node to transmit. Let this range be termed as $\theta_i \epsilon (0,1)$. As we want the collision to be as limited as it can be, we set

$$p_i^{tr}(t)\left(\mu_n^{x,y}(t)\right) \leq \theta_i \quad (8)$$

 That is a node senses the channel falsely to be idle and transmits causing a collision should be limited by θ_i. Therefore a node transmits with the following probability.

$$p_i^{tr}(t) = \min\left(1, \frac{\theta_i}{\mu_n^{x,y}(t)}\right) \quad (9)$$

Equation (9) shows the transmission being done in opportunistic or a deterministic way. If the value $\frac{\theta_i}{\mu_n^{x,y}(t)}$ is larger than 1, then the node will deterministically transmit with a probability of 1, otherwise opportunistically with a value of $\frac{\theta_i}{\mu_n^{x,y}(t)}$. Assuming the successful and unsuccessful attempts of packet transmission as Bernoulli trials, we can write the successful transmission probability after k attempts as

$$P_{suc} = p^{k-1} \times (1 - p) \quad (10)$$

Hence finally we can write a metric that incorporate the total number of transmission attempts required to successfully transmit a packet from node x to node y. we call this metric as E_{STX} and it inherently incorporates the link layer failures at a particular node.

$$E_{STX} = \sum_{k=1}^{\infty} k \times P_{suc} \quad (11)$$

The value of E_{STX} describes the characteristics achieved on a single link for a particular channel. However there can be multiple channels that are common between two nodes. So we define another metric that depicts the combined effect of all the common channels on a link between two nodes xand y. Let

$$k \in \text{set of working channels on node x}$$

$$l \in \text{set of working channels on node y}$$

Therefore, we are interested in the channels that are common between the communicating nodes x and y and define the variable i as

$$i: (i \subseteq \Psi) \text{ and } (i \in (k \cap l))$$

Therefore the combined effect over the common channels is defined as

$$E_{RANK}^{x \leftarrow y} = \sum_i E_{STX} \quad (12)$$

The metric E_{RANK}^{xy} has the property of bi-directionality and. We call this as the rank given by a node to in the backward direction neighbor.

Algorithm 1: Instantaneous Rank Calculation

1: Input [Neighbour y, Channel List L]

2: Compute: j: $(j \subseteq \Psi)$ and $(j \in (k \cap l))$

3: Index the list with j

4: For (1 to j)

5: Compute: $\mu_n(t) \leftarrow (\varepsilon_n, \delta_n)$

6: Compute: $P \leftarrow \mu_n(t), p_n^{tr}$

7: Compute: $P_{suc} \leftarrow p, k$

9: Compute $E_{STX} \leftarrow P_{suc}, k$

10: Update E_{RANK}

11: End for

Algorithm 1 computes the E_{RANK} value for a specific neighbor when a request route discovery request comes. The algorithm computes the common data channels between the two nodes and then computes the probability of success on these channels in an iterative manner. Finally an up to date value of the E_{RANK} value is used for ranking the link with that node

Algorithm 2: Periodic Rank Calculation

1: Input [Neighborset N, Channel List L]

2: Initialize Rank = [N]

3: Index nodes with i

4: For (1 to i)

5: Compute: j: $(j \subseteq \Psi)$ and $(j \in (k \cap l))$

6: Index the list with j

7: For (1 to j)

8: Compute: $\mu_n(t) \leftarrow (\varepsilon_n, \delta_n)$

9: Compute: $P \leftarrow \mu_n(t), p_n^{tr}$

10: Compute: $P_{suc} \leftarrow p, k$

11: Compute $E_{STX} \leftarrow P_{suc}, k$

12: Update E_{RANK}

13: End for

14: Rank[i] = E_{RANK}

15: End for

Algorithm 2 computes the ranks of the complete neighbor set of a node and used for ranking the links in a periodic manner. Equation (5) is used in the algorithm for computing the $\mu_n(t)$ values between the time interval of the period and a vector set is maintained. These results can be used by any node which is unable to perform sensing in a particular time interval.

3.3. Assumptions

First, we assume that each network node contains 3 radios for data traffic. Secondly all signaling/routing information uses a common control channel that is available every time and tuned on a separate dedicated radio. Thus there are total 3+1 radios at each node. Thirdly, we assume that the link existing between two communicating node is bi-directional. Lastly, we assume that at least 40% of the secondary user communication should be succeeded on every selected link in path. Therefore, we define a threshold E_{LIMIT} equal to 0.4. Also the path establishment and channel assignment is being done for a single existing source destination pair in the network.

3.4. Route Discovery

In the routing module of our proposed framework, we use the on-demand routing mechanism to build the multiple routes for a single source destination pair. Thus we modify Dynamic Source Routing (DSR) protocol for this purpose. When source node wants to communicate with destination node and no route information is known, source node initiates ROUTE REQUEST (RREQ) with new ID and broadcasts it to its 1-Hop neighbors. This process continues till the destination node receives multiple RREQ from different routes. Then the destination node assigns channel and radio to each link of the candidate path according to Algorithm 3. After this effectiveness of each candidate path is evaluated so destination can select a maximum of 4 best routes for the source. Maximum three of them will be used for the data communication and the other path will be preserved and used as a backup route, to be discussed in detail later. Selected route information is sent to the source node via the ROUTE REPLY (RREP) packet that is unicast. The selected intermediate nodes will tune their radio and channel accordingly.

3.5. RREQ Propagation

The main objective of our routing module is to construct maximally disjoint paths. Two types of disjointness; node level and spectrum level will be ensuring to construct the multiple routes. This is to prevent some nodes being congested and to efficiently utilize the network resources. Therefore it is necessary that the destination node knows the information of all the candidate paths to ensure disjointness. Thus we use the source routing approach where each forwarding node including source append their information with the RREQ message. Furthermore intermediates nodes are not allowed to send RREP to source in case they have route information in their cache.

When the source node has some data for destination node but does not have path information. It will generate a new RREQ packet with new RREQ_ID that uniquely identify the RREQ packet. Source node appends its ID and Radio, Channel, Susceptibility (RCS) table. When an intermediate node receives RREQ it will firstly ensure that the RREQ is not duplicate by checking its ID in appending list if so then discards this RREQ. If not then appends its node ID and RCS table and forwards it. In order to discover multiple paths between source and destination pair we introduce a new forwarding approach based on our metric $E_{RANK}^{x \leftarrow y}$ .Thus each intermediate node processes the RREQ as follow:

1. Check whether there exists some common data channel with the forwarded node if no then discard the RREQ.

2. Compute the $E_{RANK}^{x \leftarrow y}$ according to Algorithm 1 and compare with the E_{LIMIT}. If the $E_{RANK}^{x \leftarrow y}$ value is below the threshold then discard RREQ packet, forward otherwise.

3. If a RREQ request packet from the same ID was processed earlier from a different forwarded neighbor, then it will compute E_{RANK} vector according to the Algorithm 2 and compare the corresponding column entries. For a greater value of $E_{RANK}^{x \leftarrow y}$ for the current request as compared to the previously processed RREQ packet, the packet will be forwarded, discarded otherwise.

4. If the E_{RANK} value for both RREQ packets will be the same then Hop count will be used as tie breaker.

5. To avoid loops the packet is discarded by the previous node from where the packet had been forwarded when received.

6. Every intermediate node in the path appends its ID and Band, Radio, Channel Susceptibility (BRCS) table. The format of the table is as follows:

Table 2: The BRCS Table

Node ID			
Band B	Susceptibility S	Neighbors N	Radio R
b_1	μ_1		
b_2	μ_2	n_1, n_2	$r_1, r_2,$
b_3	μ_3		

The table includes the nodes ID, the data channels available at the node. The destination node waits a certain period of time to collect more RREQ from different paths and apply Algorithm 3 to assign channel and radio on each path. Note that all the routes are not necessarily of equal quality and length.

Hence if a request had traversed from source node S to destination node D through hops A, B and C, following IDs would have been appended in the described sequence when the request reaches node D.

Table 3: The PREQ format

Node S	Node A	Node B	Node C

3.6. Path Selection and Channel Assignment

It selects a channel for each link along the path by considering the two factors: firstly channel is not already assigned within one hope neighbor, secondly channel with minimum susceptibility among the common channel set of link should be selected. The destination waits a certain amount of time to gather multiple PREQ's from different paths. The destination evaluates each path by choosing the channel on each link with minimum susceptibility and avoiding the selection of same channel in the interference range neighbors. Algorithm followed by destination for path selection is described in Algorithm 3.

Algorithm 3: Path Selection Algorithm

1: Input: [RREQ (m)]

2: for i: 1 to m

3: for j = n to 2 where n is number of nodes in path

4: $k = j\text{-}1$

5: $°C_{k,j} = (\text{channels}(j) \cap \text{channels}(k))$

6: $c = \arg \min c \in °C_{k,j}(\mu_l)$ $l \in °C_k$

7: while $c \in c_p \cup c_k$ where p $\in$ Neighbours of k,

8: c_p is candidate channel for p

9: $°C_{k,j} = °C_{k,j} \backslash c$

10: $c = \arg \min c \in °C_{k,j}(\mu_l)$ $l \in °C_{k,j}$

11: end while

12: $c_{k,r} = c_{k,r} \cup c$ $r \in \mathfrak{R}$ r is less expected used radio

13: end for

14: $\mathbb{R}_i = {\sum_{k=2}^{n} \mu_{i,c_k}}\big/{\text{no. of hops}}$

15: end for

After receiving sufficient number of RREQ with same RREQ ID from different path, destination node computes the candidate channel for each link along the path. Candidate channel is the common channel between two nodes k and j with minimum μ_n and also it should not be in candidate channel set of neighbors of node k as well as in its own candidate channel set. If channel with minimum μ_n is also in candidate channel set of neighbors of node k or in its own candidate channel set then remove channel from the common channel list of k and j. Again compute the channel with minimum μ_n until it satisfies both conditions. Following same procedure, destination determines the candidate channel for each node along the path. In order to get primary user susceptibility for a path, sum of susceptibility on candidate channel of each link along the path is divided by number of hops. Three paths among the discovered paths with minimum rank are used in multipath routing and another can be set as a backup path. At the end of the procedure the destination node sends a ROUTE REPLY (RREP) message along the chosen paths to set the channels on the links.

4. Conclusion

The problem that the paper focuses is to discover multiple routes in cognitive mesh networks using a different methodology than previously adopted hop count and then do channel assignment on the nodes. We have defined a probability measure as primary user susceptibility on a channel sensed by a node as the percentage of the channel being in busy state. This has been done by defining a combination of markov and asymmetric channel models. This incorporates the basic characteristics of cognitive radios. From it we used a communication success probability on a channel and used this to rank the link and define a metric that can be used to forward or drop route establishment requests in terms of a defined threshold. Hence here if channels on a link are having high success rates, the requests would be forwarded and multiple paths would be established even if we are having more hops. The request contains appended data from intermediate hops in the form of channels and the primary user susceptibility. The destination then selects path preferences by the combined effect of primary user susceptibilities on the channels in the path and avoiding inter and intra flow interferences at the same time while assigning channels to the nodes.

ACKNOWLEDGEMENTS

I would like to special thanks for Miss Saba Saifullah for making available the data and other help. This work could not be completed without her help.

REFEREENCES

[1]   JOSEPH MITOLA III AND GERALD Q. MAGUIRE. COGNITIVE RADIO:MAKING SOFTWARE RADIOS MORE PERSONAL. IEEE PERSONAL COMMUNICATIONS, 6:13–18, 1999.

[2]   Simon Haykin. Cognitive radio: Brain-empowered wireless communications. IEEE Journal Selected Areas in Communications, 23:201–220, 2005.

[3]   Ian F Akyildiz, Won Yeol Lee, Mehmet C Vuran, and Shantidev Mohanty. Cnext generation/dynamic spectrum access/cognitive radio wireless networks: A survey. Computer Networks, 50:2127–2159, 2006.

[4]   IEEE 802.11s http://www.802wirelessworld.com/.

[5]   R. Hincapie, J. Tang, G. Xue and R. Bustamante, QoS routing in wireless mesh networks with cognitive radios, Proceedings of IEEE Globecom'2008.

[6]   D.B. Johnson and D.A. Maltz, .Dynamic Source Routing in Ad Hoc Wireless Networks, In Mobile Computing, 1996.

[7]   C.E. Perkins and E.M. Royer, .Ad-Hoc On Demand Distance Vector Routing, Proceedings of IEEE WMCSA'99, 1999.

[8]   Sung-Ju Lee and Mario Gerla. Split multipath routing with maximally disjoint paths in ad hoc networks. In Proceedings of IEEE ICC, 2001.

[9]   Shiwen Mao, Dennis Bushmitch, Sathya Narayanan, and Shivendra S. Panwar, "MRTP: A Multi-Flow Realtime Transport Protocol for Ad Hoc Networks", Vehicular Technology Conference, 2003.

[10]   Badoi, C.-I, Croitoru, V, Popescu, A. "HC-IPSAG Cognitive Radio Routing Protocol: Models and Performance" in 8th International Conference on Wireless and Optical Communications Networks (WOCN), 2011.

[11]   H.Khalife, S.Ahuja, N.Malouch and M.Krunz. "Probabilistic Path Selection in Opportunistic Cognitive Radio Networks" In the proceedings of the IEEE globecom conference. Orleans, USA 2008.

[12]   A. El Masri, N. Malouch. "A Routing Strategy for Cognitive Radio Networks Using Fuzzy Logic Decisions", the First International Conference on Advances in Cognitive Radio: COCORA 2011.

[13]   R. Draves, J. Padhye, and B. Zill, "Routing in Multi-Radio, Multi-Hop Wireless Mesh Networks," ACM Annual Int'l. Conf. Mobile Comp. and Net. (MOBICOM), 2004.

[14]   G. Cheng, W. Liu, Y.Z. Li, and W.Q. Cheng. Spectrum aware on-demand routing in cognitive radio networks. In Proceedings of IEEE DySPAN, 2007.

[15]   G. Chittabrata and A. Dharma P. Channel assignment with route discovery (card) using cognitive radio in multi-channel multi-radio wireless mesh networks. In Proceedings of 1st IEEE SDR Workshop, IEEE SECON, 2006.

[16]   Wang, X. and Kwon, T.T. and Choi," A multipath routing and spectrum access (MRSA) framework for cognitive radio systems in multi-radio mesh networks" Proceedings   of the 2009 ACM workshop on Cognitive radio networks.

18

PERFORMANCE ANALYSIS FOR BANDWIDTH ALLOCATION IN IEEE 802.16 BROADBAND WIRELESS NETWORKS USING BMAP QUEUEING

Said EL KAFHALI   Abdelali EL BOUCHTI   Mohamed HANINI   and   Abdelkrim HAQIQ

Computer, Networks, Mobility and Modeling laboratory
e-NGN research group, Africa and Middle East
FST, Hassan 1st University, Settat, Morocco
{kafhalisaid, a.elbouchti, haninimohamed, ahaqiq}@gmail.com

ABSTRACT

This paper presents a performance analysis for the bandwidth allocation in IEEE 802.16 broadband wireless access (BWA) networks considering the packet-level quality-of-service (QoS) constraints. Adaptive Modulation and Coding (AMC) rate based on IEEE 802.16 standard is used to adjust the transmission rate adaptively in each frame time according to channel quality in order to obtain multi-user diversity gain. To model the arrival process and the traffic source we use the Batch Markov Arrival Process (BMAP), which enables more realistic and more accurate traffic modelling. We determine analytically different performance parameters, such as average queue length, packet dropping probability, queue throughput and average packet delay. Finally, the analytical results are validated numerically.

KEYWORDS

IEEE 802.16; Quality of Service; Bandwidth Allocation; Performance Parameters; BMAP Process; OFDMA; Queueing Theory; Adaptive Modulation and Coding.

1. INTRODUCTION

1.1.   Reference system

IEEE 802.16 standard networks accommodate the increasing user demand to enable pervasive, high-speed mobile internet access to a very large coverage area. Worldwide Interoperability for Microwave Access (WiMAX), first standardised in 2004 [1] known as IEEE 802.16, can provide broadband communications over wireless for various types of multimedia traffic, such as video streaming, VoIP, FTP etc.

WiMAX presents a very challenging multiuser communication problem [12] – many users in the same geographic area will require high on-demand data rates in a finite bandwidth, with low latency. Multiple access techniques allow different users to share the available bandwidth by allotting each user some fraction of the total system resources. Due to the diverse nature of anticipated WiMAX traffic, and the challenging aspects of the system deployment (mobility, neighboring cells, and high required bandwidth efficiency), the multiple access problems are quite complicated in WiMAX.

The IEEE 802.16 standard defines two types of operating mode for sharing the wireless medium: Point-to-Multipoint (PMP) and Mesh. The PMP mode adopts a cellular architecture, in this mode subscriber stations are scattered in the cellule around a central base station. There are two directions: Downlink (from BS to SS) and Uplink (from SS to BS). Transmissions from

SSs are directed to and coordinated by the BS. On the other hand, in Mesh mode, the nodes are organized ad hoc and scheduling is distributed among them.

The WiMAX standard [1] defines the physical layer specifications and the Medium Access Control (MAC) signaling mechanisms. IEEE 802.16 uses two types of the modulation systems: OFDM (Orthogonal Frequency Division Multiple) and OFDMA (Orthogonal Frequency Division Multiple Access). OFDMA [12], extended OFDM, to accommodate many users in the same channel at the same time, and it has been adopted as the physical layer transmission technology for IEEE 802.16 based broadband wireless networks.

1.2.   Related works

In order to promise the quality of real-time traffic and allow more transmission opportunity for other traffic types, an Adaptive Bandwidth Allocation model (ABA) for multiple traffic classes in IEEE 802.16 worldwide interoperability for microwave access networks was studied in [17]. The aim of work in [28] is to show how to exploit adaptive bandwidth allocation to increase system utilization (for the system administrator) with controlled QoS degradation (for the users). Instead of only focusing on bandwidth utilization or blocking/dropping probability, two new user-perceived QoS metrics, degradation ratio and upgrade/degrade frequency, are proposed. A Markov model is then provided to derive these QoS metrics. Using this model, authors evaluate the effects of adaptive bandwidth allocation on user-perceived QoS and show the existence of trade-offs between system performance and user-perceived QoS. Mathematical tools were used in [7, 8 and 9] to study performances parameters of both the connection-level and the packet-level for a model using two Connection Admission Control (CAC) schemes considered at a subscriber station in a single-cell IEEE 802.16 environment in which the base station allocates sub-channels to the subscriber stations in its coverage area.

For wireless mobile networks, the problem of providing packet-level QoS was studied quite extensively in the literature. A scheduling mechanism for downlink transmission was proposed in [32] to provide delay guarantee. In [15], authors proposed two credits based scheduling schemes which can efficiently serve real time burst traffic with reduced latency. The effect of the proposed schemes on latency, bandwidth utilization and throughput for real time burst flows is compared with Round Robin scheduling scheme. In [4], the proposed intergraded model can be applied to IEEE.16e. This model supports quality of service for request mechanism and data transmission in the uplink phase in the presence of channel noise; the authors calculate the performance parameters for single and multichannel wireless networks, like the requests throughput, data throughput and the requests acceptance probability and data acceptance probability. In [33], a dynamic fair resource allocation scheme was proposed to support real-time and non-real-time traffic in cellular CDMA networks. In [34], authors considered a data transmission system over a wireless channel, where packets are queued at the transmitter. Using A Markov approximation, they studied the statistics of the packet dropping process due to buffer overflow under automatic repeat request (ARQ) based error control scheme.

In [27], the authors consider a point-to-point wireless transmission where link layer ARQ is used to counteract channel impairments. They presented an analytical model framework to compute link-layer packet delivery delay statistics as a function of the packet error rate. An adaptive cross-layer scheduler was proposed in [14] for multiclass data services in wireless networks. The proposed scheduler uses the queuing information as well as it takes the physical layer parameters into account so that the required QoS performances can be achieved. The capacity of TDMA and CDMA-based broadband cellular wireless systems was derived in [30] under constrained packet-level QoS.

In [13], an analytical model is proposed to study the impacts of the channel access parameters, bandwidth configuration and piggyback policy on the performance. The impacts of physical burst profile and non-saturated traffic have also been taken into account. It is observed by simulations that the bandwidth utilization can be improved if the bandwidth for random channel access can be properly configured according to the channel access parameters, piggyback policy and network traffic. Besides, there isn't a single set of configurations that is always the best for all the network scenarios.

The authors in [5] present a pipeline approach to grant bandwidth at the BS of an IEEE 802.16 FDD network with half-duplex SSs. Based on this, they proposed a grant allocation algorithm, namely, the Half-Duplex Allocation (HDA) algorithm, which always produces a feasible grant allocation provided that the sufficient conditions are met. Although there have been several proposals for QoS scheduling frameworks and algorithms in IEEE 802.16 BWA networks in the literature [24, 31], they mainly focus on the QoS architecture and scheduling algorithm in a base station to satisfy diverse QoS requirements, rather than bandwidth request algorithm in a subscriber station.

A previous researcher in an attempt to address bandwidth allocation in IEEE 802.16 was reported by the authors in [10]. They considered a similar model in OFDMA based-WiMAX but they modeled packet-level by MMPP process and they compared various QoS measures.

Since the introduction of Batch Markovian Arrival Process (BMAP) by Lucantoni [11], the researchers [16, 21, and 29] prove that BMAP enables more realistic and more accurate traffic modeling; it can also capture dependency in traffic processes and outperforms MMPP and Poisson traffic models.

Since the incoming traffic in IEEE 802.16 has a self-similarity and a bursting nature causing correlation in inter-arrival times -which influences the performance of the system- we are motivated for using BMAP which can model such traffic correlation.

1.3.   Aims of the paper

In this paper, we present a performance analysis for bandwidth allocation in IEEE 802.16 broadband wireless access networks considering the packet-level quality-of-service (QoS). Adaptive modulation and coding (AMC) rate based on IEEE 802.16 standard is used to adjust the transmission rate adaptively in each frame time according to channel quality in order to obtain multi-user diversity gain. A queueing analytical model is developed based on a Discrete-Time Markov Chain (DTMC) which captures the system dynamics in terms of the number of packets in the queue. We assume that the arrival process is modelled by the Batch Markov Arrival Process (BMAP) as the traffic source. Based on this model, various performance parameters such as average queue length, packet dropping probability due to lack of buffer space, the queue throughput, and the average queueing delay are obtained. Finally, the analytical results are validated by numerical results.

1.4.   Organisation of the paper

The rest of the paper is organized as follows: In Section 2, we briefly introduce QoS architecture of IEEE 802.16 networks. Section 3 presents Modulation and Coding Schemes for IEEE 802.16. Section 4 describes the system model. The formulation of the analytical model is presented in Section 5. In section 6, different performance parameters are analytically determined. Section 7 states numerical results. Finally, section 8 gives a conclusion of this paper.

2. QOS ARCHITECTURE OF IEEE 802.16 NETWORKS

In this paper, we consider a point-to-point wireless mode (PMP) of IEEE 802.16, where a base station (BS) serves a set of subscriber stations (SSs). The Uplink and the downlink are served in the separate region of physical layer (OFDMA/TDD) frame .the downlink channel is in broadcast mode, but an SS is only required to process data which are addressed to itself. In the uplink sub-frame, on the other hand, the SSs transmit data to the BS in a Time Division Multiple Access (TDMA) manner. Downlink and uplink sub-frames are duplexed using one of the following techniques: (Frequency Division Duplex FDD), where downlink and uplink sub-frames occur simultaneously on separate frequencies, and Time Division Duplex (TDD), where downlink and uplink sub-frames occur at different times and usually share the same frequency. SSs can be either full duplex or half-duplex.

IEEE 802.16e uses a connection-oriented medium access control (MAC) protocol which provides a mechanism for the SSs to request bandwidth to the BS. IEEE 802.16e MAC supports two classes of SS: grant per connection (GPC) and grant per SS (GPSS). In the case of GPC, bandwidth is granted to a connection individually. In contrast, for GPSS, a portion of the available bandwidth is granted to each of the SSs and each SS is responsible for allocating bandwidth among the corresponding connections.

The lengths of the downlink and uplink sub-frames for each SS are determined by the BS and broadcast to the SSs through downlink and uplink map messages (UL-MAP and DL-MAP) at the beginning of each frame. Therefore, each SS knows when and how long to receive from and transmit data to the BS. In the uplink direction, each SS can request bandwidth to the BS by using BW-request packets.

WiMAX is associated with the IEEE 802.16 standard [1, 2, and 3], which defines five classes of traffic flows representing different types of services in the following order: Unsolicited Grant Service (UGS), Extended Real Time Polling Service (ertPS), Real Time Polling Service (rtPS), Non-Real Time Polling Service (nrtPS), and Best Effort Service (BE). Each class has its QoS mechanisms at the Media Access Control (MAC) layer to support the various applications. UGS is designed to support real-time service flows that generate fixed-size data packets on a periodic basis, such as VoIP without silence suppression. ertPS supports real-time applications which generate variable-sized data packets periodically that require guaranteed data rate and delay with silence suppression. rtPS supports real-time service flows that generate variable data packets size on a periodic basis. nrtPS supports delay-tolerant data streams which are more bursty in nature, such as FTP, in general, the nrtPS can tolerate longer delays and is insensitive to delay jitter, but requires a minimum throughput. BE supports traffic with no QoS requirements, such as email, and therefore may be handled on a resource-available basis.

3. MODULATION AND CODING SCHEMES FOR IEEE 802.16

Adaptive modulation and coding scheme (AMC) is supported in the WiMAX networks. The basic idea of AMC is to maximize the data rates by adjusting the transmission parameters according to the fluctuations in the channel.

The channel quality is determined by the instantaneous received Signal-to-Noise Ratio (SNR) γ in each time slot. We assume that the channel is stationary over the transmission frame time. Lower data rates are achieved by using Modulation Level and rate error correcting corresponding to Rate $ID = 0$ (BPSK and 1/2). The higher data rates are achieved by using Modulation Level and rate error correcting corresponding to Rate $ID = 6$ (64QAM and 3/4). In all, there are 52 different possible configurations of modulation order and coding types and rates [12], although most implementations of WiMAX will offer only a fraction of these. Table 1 lists these schemes represented by different rate IDs for IEEE 802.16 WiMAX Networks.

In an OFDMA system [12], each user will be allocated a block of subcarriers, each of which will have a different set of SNR. Therefore, care needs to be paid to which constellation/coding set is chosen based on the varying SNR across the subcarriers.

To determine the mode of transmission (i.e., modulation level and coding rate), an estimated value of SNR at the receiver is used. In this case, the SNR at the receiver is divided into $N + 1$

nonoverlapping intervals (i.e., $N = 7$ in WiMAX) by thresholds $\Gamma_n (n \in \{0,1,...,N\})$ where $\Gamma_0 < \Gamma_1 < ... \Gamma_{N+1} = \infty$. The subchannel is said to be in state n (i.e., *rate ID = n* will be used) if $\Gamma_n \leq \gamma < \Gamma_{n+1}$. To avoid possible transmission error, no packet is transmitted when $\gamma < \Gamma_0$. Note that, these thresholds correspond to the required SNR specified in the WiMAX standard, SNR, Signal-to-Noise Ratio [18, 23, and 25].

That is, $\Gamma_0 = 6.4$, $\Gamma_1 = 9.4,...,$ $\Gamma_N = 24.4$ (as shown in Table 1).

Table 1: IEEE 802.16 Profiles.

Rate ID	Modulation Level (Coding)	Information Bits/Symbol	Required SNR (db)
0	BPSK (1/2)	0.5	6.4
1	QPSK (1/2)	1	9.4
2	QPSK (3/4)	1.5	11.2
3	16QAM (1/2)	2	16.4
4	16QAM (3/4)	3	18.2
5	64QAM (2/3)	4	22.7
6	64QAM (3/4)	4.5	24.4

4. MODEL DESCRIPTION

4.1. Arrival Process Traffic

The BMAP has received considerable interest during the last few years. It was first introduced by Neuts [11] as the versatile Markovian point Process. It generalizes Markovian Arrival Process (MAP) introduced by Lucantoni et al. [20].

To capture the arrival process traffic, we use a BMAP process [6]. The arrivals in the BMAP is directed by the irreducible continuous time Markov chain CTMC with a finite state space {0, 1, ..., S}. Sojourn time of the CTMC in the state s has exponential distribution with parameter λ_s. After time expires, with probability $p_0(s,s')$ the chain jumps into the state s' without generation of packets and with probability $p_k(s,s')$ the chain jumps into the state s' and a batch consisting of k packets is generated, $k \geq 1$. The introduced probabilities satisfy conditions: $p_0(s,s) = 0$, the sum of the probabilities of all outgoing transitions has to be equal to 1,

$$\sum_{k=1}^{\infty}\sum_{s'=0}^{S} p_k(s,s') + \sum_{\substack{s'=0 \\ s' \neq s}}^{S} p_k(s,s') = 1, \ 0 \leq s \leq S. \tag{1}$$

The BMAP is a two dimensional Markov process $\{A(t), J(t)\}$ on the state space $\{(i, j) / i \geq 0, 0 \leq j \leq S\}$ with infinitesimal generator given by:

$$\Psi = \begin{pmatrix} D_0 & D_1 & D_2 & D_3 & \cdots \\ 0 & D_0 & D_1 & D_2 & \cdots \\ 0 & 0 & D_0 & D_1 & \cdots \\ 0 & 0 & 0 & D_0 & \cdots \\ \vdots & \vdots & \vdots & \vdots & \ddots \end{pmatrix} \tag{2}$$

where the matrices $D_0 = [D_{ss'}]$, $0 \leq s \leq S, 0 \leq s' \leq S$ has negative diagonal elements and non negative off diagonal elements given by:

$$D_{ss'} = \begin{cases} -\lambda_s, & s' = s \\ \lambda_s p(s,s'), & s \neq s' \end{cases} \tag{3}$$

The matrices $D_k, k > 0$ are defined by:

$$D_k = [D_{k,ss'}], \ 0 \leq s \leq S, 0 \leq s' \leq S, \ k > 0 \tag{4}$$

where: $\quad D_{k,ss'} = \lambda_s p_k(s,s'), \ 0 \leq s \leq S, 0 \leq s' \leq S, \ k > 0. \tag{5}$

The matrix D defined by. $D = \sum_{k=0}^{\infty} D_k$, is an irreducible infinitesimal generator. We also

assume that $D_k \neq D_0$, which ensures that arrivals will occur.

The variable $A(t)$ counts the number of arrivals during $[0,t[$ and the variable $J(t)$ represents the phase of the arrivals process.

The steady-state probability vector π_{BMAP} of the CTMC with generator D can be calculated as usual:

$$\pi_{BMAP}.D = \vec{0}, \quad \pi_{BMAP}.e = 1. \tag{6}$$

Where $\vec{0}$ and e are row and column vectors consisting of zeros and units, respectively.
The mean steady-state arrival rate generated by the BMAP is:

$$\lambda_{BMAP} = \pi_{BMAP} \sum_{k=1}^{\infty} k D_k e. \tag{7}$$

More detail and results concerning this Process can be found in [19] for instance.
The probability $f_a(\lambda_s, T)$ of $a = 0,1,...,A$ incoming packets, with A denoting the maximum

packets' number, that arrive with mean rate λ_s within a time slot interval T is given by:

$$f_a(\lambda_s, T) = \frac{e^{-\lambda_s T}(\lambda_s T)^a}{a!} \tag{8}$$

It is also essential for the condition $\sum_{a=A}^{+\infty} \frac{e^{-\lambda_s T}(\lambda_s T)^a}{a!} < er \ \forall s$ always to stand, with er expressing

a sufficiently small number.
Note that the probability that a Poison arrivals with average rate λ_s occur during an interval T

is given by the $S \times S$ diagonal matrix ξ_a which is defined as:

$$\xi_a = \begin{bmatrix} f_a(\lambda_1, T) & & & \\ & f_a(\lambda_2, T) & & \\ & & \ddots & \\ & & & f_a(\lambda_s, T) \end{bmatrix} \tag{9}$$

4.2. System Model

We consider an infrastructure-based wireless access network, where connections are established between a base station (BS) and multiple subscribers stations (SSs) through a TDMA/TDD access mode using single carrier air-interface (as shown in Figure 1). Each subscriber station serves multiple connections. For each connection a separate queue in SS with size X packets is used for buffering the packets from higher layers. In particular, for one connection, there is a queue for uplink and another queue for downlink transmissions from the SS and the BS, respectively. We consider an SS of type GPC. Therefore, through the SS a certain amount of bandwidth is reserved for each connection during bandwidth allocation.

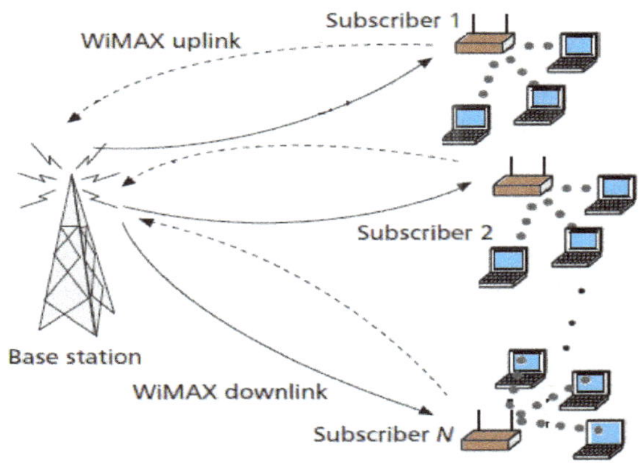

Figure 1: System model

5. FORMULATION OF THE ANALYTICAL MODEL

5.1. State Space

The state of the queue is observed at the beginning of each frame. We assume that connection i is allocated with b_i units of bandwidth and a packets arriving during frame period f will not be transmitted until frame period $f+1$ at the earliest. The state space of the queue can be defined as follows:

$$E = \{(x,s); 0 \leq x \leq X, 1 \leq s \leq S\}, \tag{10}$$

where x, s represent, the number of packets in the queue, the state (phase) of an irreducible continuous time Markov chain of BMAP arrival process respectively.

5.2. Transition Matrix for the Queue

The transition matrix M for the queue can be expressed as follows:

$$M = \begin{bmatrix} m_{0,0} & m_{0,1} & \cdots & m_{0,A} & & & \\ \vdots & \vdots & \ddots & & \ddots & & \\ m_{D,0} & m_{D,1} & \cdots & m_{D,D} & \cdots & \cdots & m_{D,D+A} \\ \ddots & & \ddots & \ddots & \ddots & \ddots & & \ddots \\ & & m_{X-A,X-A-D} & \cdots & \cdots & & m_{X-A,X} \\ & & \ddots & \ddots & \ddots & \ddots & \vdots \\ & & & m_{X,X-D} & \cdots & \cdots & m_{X,X} \end{bmatrix} \tag{11}$$

The rows of matrix M represent the number of packets in the queue and element $m_{x,x'}$ inside this matrix denotes the transition probability for the case when the number of packets in the queue changes from x in the current frame to x' in the next frame. Also, the maximum number of packets that can enter into or depart from the queue within a frame time is represented with A and D respectively.

The maximum number of transmitted packets within a time slot is given by $D' = \min(x, D)$. Hence, if k represents the number of the successfully transmitted packets. The probability that k packets will be transmitted in a timeslot is obtained by the matrix T_k which is defined as follows:

$$T_k = \begin{cases} \binom{x}{k}(1-\theta)^k \theta^{x-k} & , k < D' \\ \sum_{j=U}^{x} \binom{x}{j}(1-\theta)^j \theta^{x-j} & , k = D' \end{cases} \tag{12}$$

where θ is the probability that a packets is successfully transmitted.

The elements in the matrix M can be obtained as follows:

$$m_{x,x-u} = \Psi \times \sum_{k-a=u} \xi_a \times T_k \tag{13}$$

$$m_{x,x+v} = \Psi \times \sum_{a-k=v} \xi_a \times T_k \tag{14}$$

$$m_{x,x} = \Psi \times \sum_{k=a} \xi_a \times T_k \tag{15}$$

for $u = 1, 2, ..., D'$ and $v = 1, 2, ..., A$ where, $k \in \{0, 1, ..., D'\}$ and $a \in \{0, 1, ..., A\}$ represent the number of departed packets and the number of packets arrivals, respectively.

With $m_{x,x-u}, m_{x,x+v}$ and $m_{x,x}$ we represent the probability that the number of packets in the queue increases by u, decreases by v, and does not change, respectively.

The remaining rows $\{x = X-A+1, X-A+2, ..., X\}$ of the matrix M, include the occurrence where some packets would be dropped due to lack of queue space. We calculate the probabilities as follows:

$$m_{x,x+v} = \sum_{a=v}^{A} m'_{x,x+a} \quad \text{for } x+v \geq X \tag{16}$$

Which express that no packet has been dropped and occur when more incoming packets to an already fully queue are dropped. Additionally, the last element of the main diagonal of M is given by:

$$m_{x,x} = m'_{x,x} + \sum_{a=1}^{A} m'_{x,x+a} \quad \text{for } x = X \tag{17}$$

where $m'_{x,x}$ is obtained for the case without any packets dropping.

Equations (16) and (17) indicate the case that the queue will be full if the number of incoming packets is greater than the available space in the queue. In other words, the transition probability to the state that the queue is full can be calculated as the sum of all the probabilities that make the number of packets in queue equal to or larger than the queue size X.

6. PERFORMANCE PARAMETERS

The performance parameters are analytically calculated using the steady state probability of the system. The vector π of these probabilities is obtained by solving the system $\pi.M = \pi$ and $\pi.1 = 1$, where 1 is a column matrix of ones.

The steady-state probabilities, denoted by $\pi(x,s)$ for the state that there are $x \in \{0,1,...,X\}$ packets in the queue, can be extracted from matrix π as follows:

$$\pi(x,s)=[\pi]_{X \times S} , s=1,...,S \tag{18}$$

The matrix π contains the steady state probabilities corresponding to the number of packets in the queue and the state (phase) of an irreducible continuous time Markov chain of BMAP arrival process. Using the steady state probabilities, the various performance measures can be obtained.

6.1.   Average Queue Length

The average number of packets in the transmission queue is obtained as follows:

$$\overline{X} = E(x) = \sum_{x=0}^{X} x \sum_{s=1}^{S} \pi(x,s) \tag{19}$$

6.2.   Packet Dropping Probability

In order to compute the packet dropping probability (p_{drop}), we firstly obtain the number of dropped packets per time slot, obtained using the average number of dropped packets per frame. If x is the number of packets in the queue and this number increases by n, the number of dropped packets X_{drop} is: $X_{drop} = n - (X-x).\sim_{]X-n,+\infty[}(x)$ where

$$\sim_A(x) = \begin{cases} 1 & if \ x \in A \\ 0 & otherwise \end{cases} \tag{20}$$

The average number of dropped packets per frame is obtained as follows:

$$\begin{aligned} \overline{X}_{drop} &= E(X_{drop}) \\ &= \sum_{x=0}^{X}\sum_{s=1}^{S}\sum_{n=X-x+1}^{A}\left(\sum_{l=1}^{S}[m_{x,x+n}]_{s,l}\right).(n-(X-x)).\pi(x,s) \end{aligned} \tag{21}$$

where the term $\left(\sum_{l=1}^{S}[m_{x,x+n}]_{s,l}\right)$ in Equation (21) indicates the total probability that the number of packets in the queue increases by n at every arrival phase. The probability $m_{x,x+n}$ is used rather than the probability of packet arrival, because the packet transmission in the same frame is considered.

After calculating the average number of dropped packets per frame, we can obtain the probability that an incoming packet is dropped as follows:

$$p_{drop} = \frac{\overline{X}_{drop}}{\lambda_{BMAP}} \tag{22}$$

where λ_{BMAP} is the mean steady state arrival rate generated by the BMAP (as obtained from (7)).

6.3. Queue throughput

It measures the number of packets transmitted in one frame and can be obtained from:

$$\varphi = \lambda_{BMAP}(1 - p_{drop}) \tag{23}$$

6.4.   Average Packet Delay

The average packet delay is defined as the number of frames that a packet waits in the queue since its arrival before it is transmitted.
We use Little's formula [22] to obtain average packets delay as follows:

$$\overline{D} = \frac{\overline{X}}{\varphi} = \frac{\overline{X}}{\lambda_{BMAP}(1 - P_{drop})} \tag{24}$$

where φ is the throughput and $\overline{X}$ is the average queue length.

7. NUMERICAL RESULTS

In the next, performance parameters are numerically evaluated, using Matlab software.

7.1.   Parameter Setting

We consider the system model depicted in section 4. Adaptive Modulation and Coding (AMC) is used in which the modulation level and the coding rate are increased if the channel quality permits. Table 1 lists these schemes represented by different rate IDs for IEEE 802.16.
The maximum number of packets that can be transmitted in one frame period is 150 packets per frame.
The queue size is assumed to be 150 packets (i.e. $X = 150$).
For simplicity of computation we assume the maximum size of the batch to be 2, and the matrices governing the state transitions of the BMAP are giving as follows:

$$D_0 = \begin{pmatrix} -2 & \frac{1}{2} \\ \frac{1}{8} & -1 \end{pmatrix}, D_1 = \begin{pmatrix} \frac{1}{2} & \frac{1}{4} \\ \frac{1}{4} & \frac{1}{4} \end{pmatrix}, D_2 = \begin{pmatrix} \frac{1}{4} & \frac{1}{2} \\ \frac{1}{4} & \frac{1}{8} \end{pmatrix}$$

Its sojourn times at each state are assumed to be exponentially distributed with rates $\lambda_0 = 2.0$, $\lambda_1 = 1.0$ respectively.

The performance parameters are measured respectively under different amount of allocated bandwidth b, under different channel qualities with constant traffic intensity, and under different traffic intensities with channel SNR in the range of rate $IDn = 0$.
In this work, bandwidth b is defined (as in [10]) as the number of packets that can be transmitted in one frame using rate $IDn = 0$.

7.2.   Results and Discussion

We first examine the impact of traffic intensity on bandwidth allocation. Variations in throughput with traffic intensity are shown in Figure 2. When the traffic intensity increases, the throughput increases until it becomes saturated. At this point (e.g., 3.0), the arriving packets cannot be transmitted faster than the transmission rate that the channel quality allows.

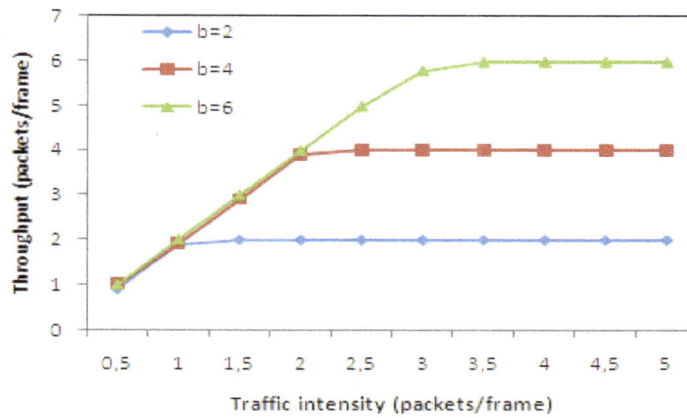

Figure 2: Throughput under traffic intensity.

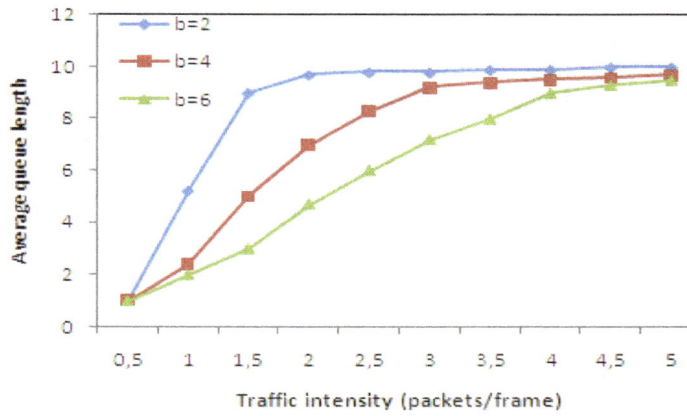

Figure 3: Average queue length under traffic intensity.

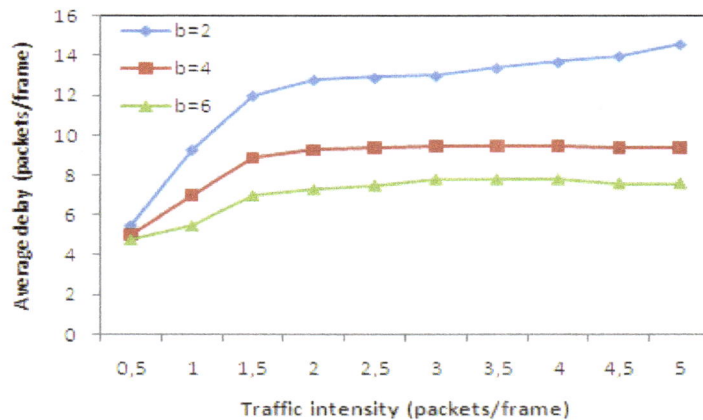

Figure 4: Average delay under traffic intensity.

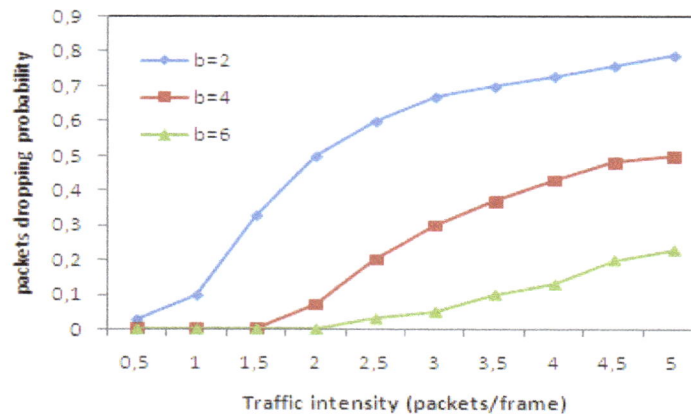

Figure 5: packets dropping probability under traffic intensity.

Average queue length, average delay and packets dropping probability increase as the traffic intensity increases (Figures 3, 4 and 5). Therefore, we can say that the increase of the parameters previously mentioned is linked with the traffic intensity. On the other hand, those parameters decrease as the channel quality improves (Figures 6, 7 and 8).

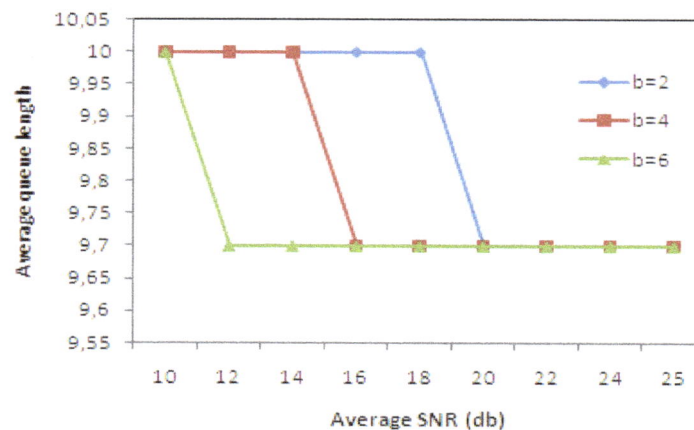

Figure 6: Average queue length under different channel qualities.

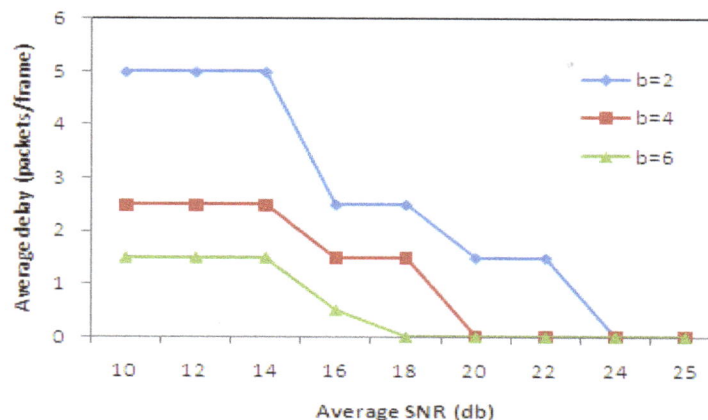

Figure 7: Average delay under different channel qualities.

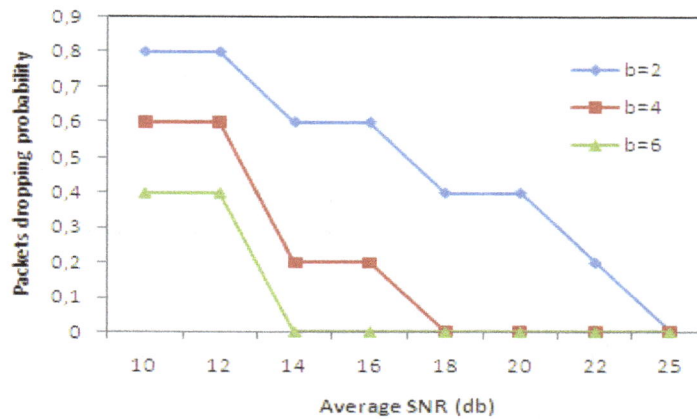

Figure 8: Packet dropping probability under different channel qualities.

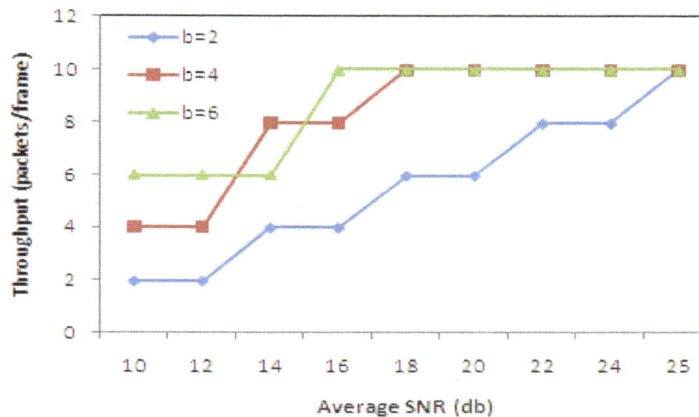

Figure 9: Throughput under different channel qualities.

When the channel quality improves, the transmitter can utilize higher modulation level and code rate to increase throughput (Figure 9). Note that if transmission rate is high enough to accommodate most of the arrival traffic, increased amount of allocated bandwidth or better channel quality will not impact the queue throughput since all the packets can be transmitted within a few frames. Moreover, different amount of allocated bandwidth results in different throughput.

8. CONCLUSION

In this paper, a queuing analytical model based on a Discrete-Time Markov Chain (DTMC) has been presented to analyze the packet-level performance in IEEE 802.16 broadband wireless access networks considering adaptive modulation and coding at the OFDMA physical layer. In the considered WiMAX system model, a base station serves multiple subscriber stations, and the base station allocates a certain number of sub-channels for each subscriber station.

To model the arrival process and the traffic sources we use the Batch Markov Arrival Process (BMAP), which enables more realistic and more accurate traffic modelling.

Using this queuing model, the impact of different traffic sources and the impact of channel quality on QoS parameters, such as average queue length, packet dropping probability, queue throughput and average packet delay, are analytically studied. Finally, the analytical results are validated numerically.

REFERENCES

[1] IEEE 802.16 WG, "IEEE standard for local and metropolitan area networks part 16: Air interface for fixed broadband wireless access systems," IEEE 802.16 Standard, June 2004.

[2] IEEE 802.16 WG, "IEEE standard for local and metropolitan area networks part 16: Air interface for fixed and mobile broadband wireless access systems, Amendment 2," IEEE 802.16 Standard, December 2005.

[3] Mobile WiMAX – Part I: A Technical Overview and Performance Evaluation, August, 2006.

[4] Abdelsalam Amer and Fayez Gebali, "General Model for Single and Multiple Channels WLANs with Quality of Service Support," International Journal of Wireless & Mobile Networks (IJWMN), Vol 1, No 2, November 2009.

[5] Andrea Bacioccola, Claudio Cicconetti, Alessandro Erta, Luciano Lenzini, and Enzo Mingozzi, Alessandro Erta, Luciano Lenzini, and Enzo Mingozzi," Bandwidth Allocation with Half-Duplex Stations in IEEE 802.16 Wireless Networks", IEEE transactions on mobile computing, vol. 6, no. 12, december 2007.

[6] B. Baynat, S. Doirieux, G. Nogueira, M. Maqbool, and M. Coupechoux, "An efficient analytical model for wimax networks with multiple traffic profiles," in Proc. of ACM/IET/ICST IWPAWN, September 2008.

[7] Abdelali EL BOUCHTI, Abdelkrim HAQIQ and Said EL KAFHALI, "Analysis of Quality of Service Performances of Connection Admission Control Mechanisms in OFDMA IEEE 802.16 Network using BMAP Queuing", International Journal of Computer Science Issues (IJCSI), Vol. 9, Issue 1, No 2, ISSN (Online): 1694-0814, pp. 302-310, January 2012.

[8] Abdelali EL BOUCHTI, Said EL KAFHALI, and Abdelkrim HAQIQ "Performance Modeling and Analysis of Connection Admission Control in OFDMA based WiMAX System with MMPP Queueing" World of Computer Science and Information Technology Journal (WCSIT), Vol. 1, No. 4, pp. 148-156 , 2011.

[9] Abdelali EL BOUCHTI, Said EL KAFHALI, and Abdelkrim HAQIQ "Performance Analysis of Connection Admission Control Scheme in IEEE 802.16 OFDMA Networks" International Journal of Computer Science and Information Security (IJCSIS), Vol. 9, No. 3, pp. 45-51 March 2011.

[10] M. Fathi, H. Taheri ,"Queuing analysis for dynamic bandwidth allocation in IEEE 802.16 standard", 3rd IEEE International symposium on wireless pervasive computing , 7-9 May 2008.

[11] M.F.Neuts,"Aversatile Markovian Point Process", J.Appl.Prob, 16:764-779, 1979.

[12] Jeffrey G. Andrews," Orthogonal Frequency Division Multiple Access (OFDMA)", book chapter, July 29, 2006.

[13] Jianhua He, Kun Yang, Ken Guild, and Hsiao-Hwa Chen, "On Bandwidth Request Mechanism with Piggyback in Fixed IEEE 802.16 Networks", IEEE transactions on wireless communications, vol. 7, no. 12, december 2008.

[14] K.B. Johnsson and D.C. Cox, "An Adaptive Cross-Layer Scheduler for Improved QoS Support of Multiclass Data Services on Wireless Systems," IEEE J. Selected Areas in Comm., vol. 23, no. 2, pp. 334- 343, Feb. 2005.

[15] C.Kalyana Chakravarthy and Prof. P.V.G.D. Prasad Reddy , "Selfless Distributed Credit Based Scheduling For Improved QoS In IEEE 802.16 WBA Networks," International Journal of Wireless & Mobile Networks (IJWMN), Vol 1, No 2, pp. 118-125, November 2009.

[16] A. Klemm, C. Lindemann, and M. Lohmann, "Traffic Modeling of IP Networks Using the Batch Markovian Arrival Process", in Proc. Computer Performance Evaluation / TOOLS, 2002, pp.92-110.

[17] T.-L. Sheu K.-C. Huang, "Adaptive bandwidth allocation model for multiple traffic classes in IEEE 802.16 worldwide interoperability for microwave access networks", The Institution of Engineering and Technology, Vol. 5, Iss. 1, pp. 90–98, 2011.

[18] Q.Liu, S. Zhou, and G. B. Giannakis, "Queuing with adaptive modulation and coding over wireless links: cross-layer analysis and design," IEEE Transactions on Wireless Communications, vol. 4, no. 2, pp. 1142–1153, May 2005.

[19] D.M. Lucantoni, "The BMAP/G/1 queue : A tutorial", In L.Donatiello and R.Nelson, editors, Performance Evaluation of Computer and Communications Systems, pages 330-358, lectures Notes in Computer Science 729, Springer Verlag, 1993.

[20] D.M.Lucantoni, K.S.Meier-Hellstern, and M.F.Neuts,"A single server queue with server vacations and a class of non-renewal arrival processes", Adv.Appl.Prob,22: 676-705,1990.

[21] D. M. Lucantoni, New Results on the Single Server Queue with a Batch Markovian Arrival Process, Comm. in Statistics: Stochastic Models 7, 1-46, 1991.

[22] R. Nelson, "Probability, stochastic process, and queueing theory", Springer-Verlag, third printing, 2000.

[23] D. Niyato and E. Hossain, "Delay-based admission control using fuzzy logic for OFDMA broadband wireless networks," in Proc. IEEE ICC'06, June 2006.

[24] D. Niyato, E. Hossain, "Queue-Aware Uplink Bandwidth Allocation and Rate Control for Polling Service in IEEE 802.16 Broadband Wireless Networks," IEEE trans. on Mobile Computing, Vol. 5, No. 6, June 2006.

[25] D. Niyato and E. Hossain, "Connection admission control in OFDMA-based WiMAX networks: Performance modeling and analysis," invited chapter in WiMax/MobileFi: Advanced Research and Technology, (Ed. Y. Xiao), Auerbach Publications, CRC Press, December 2007.

[26] D. Niyato, E. Hossain, "QoS-aware bandwidth allocation and admission control in IEEE 802.16 broad band wireless access networks: A non-cooperative game theoretic approach," Comut. Netw. (2007), doi:10.1016/jcomnet.2007.01.031.

[27] M. Rossi and M. Zorzi, "Analysis and Heuristics for the Characterization of Selective Repeat ARQ Delay Statistics over Wireless Channels," IEEE Trans. Vehicular Technology, vol. 52, no. 5, pp. 1365-1377, Sept. 2003.

[28] C.T. Chou and K.G. Shin, "Analysis of Adaptive Bandwidth Allocation in Wireless Networks with Multilevel Degradable Quality of Service," IEEE Trans. Mobile Computing, vol. 3, no. 1, pp. 5-17, Jan.-Mar. 2004.

[29] Takine, T. and Takahashi, Y. 1998. On the relationship between queue lengths at a random instant and at a departure in the stationary queue with BMAP arrivals. Stoch. Mod. 14 601–610.

[30] S.V. Krishnamurthy, A.S. Acampora, and M. Zorzi, "On the Radio Capacity of TDMA and CDMA for Broadband Wireless Packet Communications," IEEE Trans. Vehicular Technology, vol. 52, no. 1, pp. 60-70, Jan. 2003.

[31] K. Wongthavarawat and A. Ganz, "Packet scheduling for QoS support in IEEE 802.16 broadband wireless access systems," International Journal of Communication systems, vol. 16, pp. 81–96, 2003.

[32] D. Wu and R. Negi, "Downlink Scheduling in a Cellular Network for Quality-of-Service Assurance," IEEE Trans. Vehicular Technology, vol. 53, no. 5, pp. 1547-1557, Sept. 2004.

[33] L. Xu, X. Shen, and J.W. Mark, "Fair Resource Allocation with Guaranteed Statistical QoS for Multimedia Traffic in Wideband CDMA Cellular Network," IEEE Trans. Mobile Computing, vol. 4, no. 2, pp. 166-177, Mar.-Apr. 2005.

[34] M. Zorzi, "Packet Dropping Statistics of a Data-Link Protocol for Wireless Local Communications," IEEE Trans. Vehicular Technology, vol. 52, no. 1, pp. 71-79, Jan. 2003.

Permissions

All chapters in this book were first published in IJWMN, by AIRCC Publishing Corporation; hereby published with permission under the Creative Commons Attribution License or equivalent. Every chapter published in this book has been scrutinized by our experts. Their significance has been extensively debated. The topics covered herein carry significant findings which will fuel the growth of the discipline. They may even be implemented as practical applications or may be referred to as a beginning point for another development.

The contributors of this book come from diverse backgrounds, making this book a truly international effort. This book will bring forth new frontiers with its revolutionizing research information and detailed analysis of the nascent developments around the world.

We would like to thank all the contributing authors for lending their expertise to make the book truly unique. They have played a crucial role in the development of this book. Without their invaluable contributions this book wouldn't have been possible. They have made vital efforts to compile up to date information on the varied aspects of this subject to make this book a valuable addition to the collection of many professionals and students.

This book was conceptualized with the vision of imparting up-to-date information and advanced data in this field. To ensure the same, a matchless editorial board was set up. Every individual on the board went through rigorous rounds of assessment to prove their worth. After which they invested a large part of their time researching and compiling the most relevant data for our readers.

The editorial board has been involved in producing this book since its inception. They have spent rigorous hours researching and exploring the diverse topics which have resulted in the successful publishing of this book. They have passed on their knowledge of decades through this book. To expedite this challenging task, the publisher supported the team at every step. A small team of assistant editors was also appointed to further simplify the editing procedure and attain best results for the readers.

Apart from the editorial board, the designing team has also invested a significant amount of their time in understanding the subject and creating the most relevant covers. They scrutinized every image to scout for the most suitable representation of the subject and create an appropriate cover for the book.

The publishing team has been an ardent support to the editorial, designing and production team. Their endless efforts to recruit the best for this project, has resulted in the accomplishment of this book. They are a veteran in the field of academics and their pool of knowledge is as vast as their experience in printing. Their expertise and guidance has proved useful at every step. Their uncompromising quality standards have made this book an exceptional effort. Their encouragement from time to time has been an inspiration for everyone.

The publisher and the editorial board hope that this book will prove to be a valuable piece of knowledge for researchers, students, practitioners and scholars across the globe.

List of Contributors

Mst. Nargis Aktar
Department of Information and Communication Technology Mawlana Bhashani Science and Technology University, Bangladesh

Muhammad Shahin Uddin
Department of Electronics, Kookmin University, Seoul, South Korea

Md. Ruhul Amin
Department of Electrical and Electronic Engineering Islamic University of Technology, Dhaka, Bangladesh

Md. Mortuza Ali
Department of Electrical and Electronic Engineering Rajshahi University of Engineering and Technology, Bangladesh

Navneet Kaur
Department of Electronics and Communication Engineering Lovely Professional University, Phagwara Punjab, India

Lavish Kansal
Department of Electronics and Communication Engineering Lovely Professional University, Phagwara Punjab, India

Navjot Kaur
Lovely Professional University, Phagwara

Lavish Kansal
Lovely Professional University, Phagwara

Dhananjay Bisen
M.Tech, School Of Information Technology, RGPV, BHOPAL, INDIA

Sanjeev Sharma
Reader & Head, School Of Information Technology, RGPV, BHOPAL, INDIA

Debdutta Barman Roy
Department of Information Technology, Calcutta Institute of Eng. & Mgmt, Kolkata, India

Rituparna Chaki
Department of Computer Science and Eng., West Bengal University of Technology, Kolkata, India

Moez Hizem
Innov'Com Laboratory, Sup'Com, University of Carthage, Tunis, Tunisia

Ridha Bouallegue
Innov'Com Laboratory, Sup'Com, University of Carthage, Tunis, Tunisia

Koushik Majumder
Department of Computer Science & Engineering, West Bengal University of Technology, Kolkata, INDIA

Sudhabindu Ray
Department of Electronics and Telecommunication Engineering, Jadavpur University, Kolkata, INDIA

Subir Kumar Sarkar
Department of Electronics and Telecommunication Engineering, Jadavpur University, Kolkata, INDIA

E. Martin
Department of Electrical Engineering and Computer Science University of California, Berkeley California, USA

R. Bajcsy
Department of Electrical Engineering and Computer Science University of California, Berkeley California, USA

G.Fathima
Adhiyamaan College of Engineering, Hosur, TamilNadu, India

R.S.D.Wahidabanu
Govt. College of Engineering, Salem, TamilNadu, India

Abdelhakim Khlifi
National Engineering School of Tunis, Tunisia

Ridha Bouallegue
Sup'Com, Tunisia

Shumon Alam
Center of Excellence for Communication Systems Technology Research Department of Electrical and Computer Engineering Prairie View A & M University, TX 77446 United States of America

O. Olabiyi
Center of Excellence for Communication Systems Technology Research Department of Electrical and Computer Engineering Prairie View A & M University, TX 77446 United States of America

O. Odejide
Center of Excellence for Communication Systems Technology Research Department of Electrical and Computer Engineering Prairie View A & M University, TX 77446 United States of America

A. Annamalai
Center of Excellence for Communication Systems Technology Research Department of Electrical and Computer Engineering Prairie View A & M University, TX 77446 United States of America

Priyanka Mishra
Department of Electronics Engineering, United Group of Institution, Allahabad

Rahul Vij
Department of Electronics Engineering, L R of Institute of Engineering, Solan

Gurpreet Singh
Department of Electronics Engineering, Shaheed Bhagat Singh State Technical Campus, Ferozpur, Punjab

Gaurav Chandil
Department of Electronics Engineering, United Group of Institution, Allahabad

Monir Hossen and Masanori Hanawa
Interdisciplinary Graduate School of Medicine and Engineering University of Yamanashi, Japan

Lachhman Das Dhomeja
Institute of Information and Communication Technology, University of Sindh, Jamshoro, Pakistan

Shazia Abbasi
Institute of Information and Communication Technology, University of Sindh, Jamshoro, Pakistan

Asad Ali Shaikh
Institute of Information and Communication Technology, University of Sindh, Jamshoro, Pakistan

Yasir Arfat Malkani
Institute of Mathematics and Computer Science, University of Sindh, Jamshoro, Pakistan

Tony Tsang
Department of Computer Engineering, La Trobe University, Melbourne, Australia

Farhana Afroz
Faculty of Engineering and Information Technology, University of Technology, Sydney, Australia

Shouman Barua
Kumbesan Sandrasegaran

Amjad Ali
School of Electrical Engineering and Computer Sciences National University of Science and Technology, Pakistan

Muddesar Iqbal
Faculty of Computer Science & Information Technology, University of Gujrat, Pakistan

Adeel Baig
School of Electrical Engineering and Computer Sciences National University of Science and Technology, Pakistan

Xingheng Wang
College of Engineering, Swansea University, Swansea, UK

Said EL KAFHALI Abdelali EL BOUCHTI Mohamed HANINI
Computer, Networks, Mobility and Modeling laboratory e-NGN research group, Africa and Middle East FST, Hassan 1st University, Settat, Morocco

Abdelkrim HAQIQ
Computer, Networks, Mobility and Modeling laboratory e-NGN research group, Africa and Middle East FST, Hassan 1st University, Settat, Morocco